# SPACE
# MISSION PATCHES

ACC ART BOOKS

# Contents

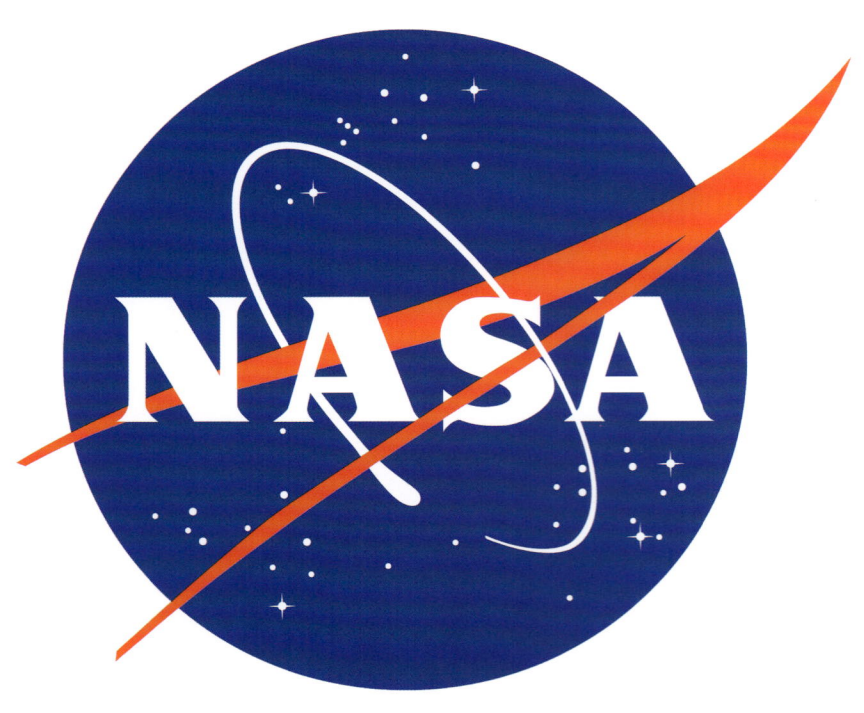

| | |
|---|---|
| **Introduction** | 3 |
| **1. A Series of Firsts** | **4** |
| Project Mercury 1958-63 | 6 |
| Project Gemini 1961-66 | 8 |
| Project Apollo 1961-72 | 12 |
| Skylab 1973 | 16 |
| Apollo-Soyuz 1975 | 17 |
| **2. Space Shuttles and Their Missions** | **18** |
| Space Shuttle Columbia 1981-2003 | 20 |
| Space Shuttle Challenger 1983-86 | 26 |
| Space Shuttle Discovery 1984-2011 | 30 |
| Space Shuttle Atlantis 1984-2011 | 38 |
| Space Shuttle Endeavour 1992-2011 | 46 |
| **3. International Space Station** | **52** |
| **4. The Future of Spaceflight** | **60** |
| SpaceX 2020-ongoing | 62 |
| Artemis Programme 2017-ongoing | 63 |

# Introduction

*Space: The Mission Patches* is an artistic time capsule representing NASA's historic missions, from Mercury to the Artemis programme. For each mission to the Moon, the International Space Station or a far corner of the solar system, NASA's tradition is to create a distinctive mission patch to symbolise and commemorate the endeavour.

The very first mission patch was introduced in 1965 by Air Force pilot and astronaut Gordon Cooper to represent his upcoming Gemini 5 mission. This was followed by hundreds of imaginative patches, including for the Apollo missions, the Space Shuttle missions and the International Space Station crews, as well as for the Artemis missions to the Moon and Mars, all of which are included within this book. These mission patches also reflect the varying artistic styles and cultural persuasions of the 1960s onward.

In addition to the mission patches, NASA's famed 'Meatball' and 'Worm' logos are among the world's most recognised and popular logos; appealing to people the world over. When Richard Danne and Bruce Blackburn designed the 'Worm' logo in 1974, they could never have imagined their logo would continue to gain in popularity as generation after generation have embraced it.

NASA's mission patches are the small artistic expressions that remind us of the benefits of collaboration on a global scale. This book will enable readers to thoroughly enjoy the visual aspects of these historic insignias, and, just maybe, begin to explore the deeper meaning of each mission.
**Bill Schwartz**

*'Back in 1974, my partner Bruce Blackburn and I created the NASA logotype, affectionately known as "The Worm". Though we always strive for timeless design, we could not have known this iconic logo would be more beloved and globally popular today than 50 years ago. Powerful in its simplicity, it still speaks to technology, innovation, and the future. "The Worm", the "Meatball" and [the] mission patches in this book clearly demonstrate NASA's enduring spirit of imagination and exploration.'* **Richard Danne**

# Chapter 1: A Series of Firsts

*Mercury, Gemini, Apollo, Skylab, Apollo-Soyuz*

| | |
|---|---|
| **5 May 1961 - Mercury 3** | Astronaut Alan Shepard successfully piloted the USA's first human spaceflight. This was followed shortly after by Virgil 'Gus' Grissom's Mercury 4 flight (1 July 1961). |
| **20 February 1962 - Mercury 6** | The first American in orbit, astronaut John Glenn circled the Earth three times in just under five hours. |
| **3 June 1965 - Gemini 4** | Astronaut Edward H. White became the first American to perform a spacewalk. |
| **27 January 1967** | One of the darkest days in NASA's history. A fire on the launch pad killed the crew of Apollo 1. The astronauts in question were Ed White, Virgil 'Gus' Grissom and Roger Chaffee. |
| **24 December (Christmas Eve) 1968** | The first people to orbit the Moon, the crew of Apollo 8 were returning from the dark side of the Moon when they saw the Earth emerging over the lunar horizon and took the world-changing photograph that would become known as *Earthrise*. |
| **20 July 1969 - Apollo 11** | Astronauts Neil Armstrong and Edwin E. (Buzz) Aldrin became the first humans to walk on the Moon. |
| **13 April 1970** | The unlucky Apollo 13 flight was afflicted with a torrent of malfunctions after an oxygen tank in the Odyssey command module ignited and blew up, prompting the crew to contact mission control with the haunting distress call: 'Houston, we've had a problem'. |
| **14 December 1972 - Apollo 17** | On this date, the final human visitors to the Moon took off from the lunar surface to return to Earth. |
| **14 May 1973** | Skylab launched, becoming the first US space station. |
| **17 July 1975** | Apollo-Soyuz linkup. The first international spaceflight. The American Apollo and Russian Soyuz crafts joined together in Earth orbit. |

# Project Mercury 1958-63

This American spaceflight programme set out to be the first to put a human into Earth's orbit and return them safely.

## 5 May 1961
## Mercury-Redstone 3: Freedom 7

First US human spaceflight with the objective of putting a human into orbit around the Earth to see how they withstand the g-forces of launching into space and re-entering the atmosphere. Alan Shepard piloted the space capsule Freedom 7 for this mission, which took 15 minutes and 28 seconds.

## 21 Jul 1961
## Mercury-Redstone 4: Liberty Bell 7

Second US human spaceflight, with the same objective as the first. Astronaut Virgil 'Gus' Grissom piloted the space capsule Liberty Bell 7 in a flight that lasted 15 mins and 30 secs.

## 20 Feb 1962
## Mercury-Atlas 6: Friendship 7

First US orbital human spaceflight, piloted by John Glenn. The spacecraft flew on a trajectory that would enable it to stay in space and complete at least one full orbit of the Earth. Mission time was four hours and 55 mins.

## 24 May 1962
## Mercury-Atlas 7: Aurora 7

In a repeat of Glenn's mission three months prior, Scott Carpenter became the sixth human to fly in space. His mission was to fly his space capsule Aurora 7 for three Earth orbits. Duration was four hours, 56 mins and 5 secs.

## 3 Oct 1962
## Mercury-Atlas 8: Sigma 7

This fifth US human spaceflight was America's longest crewed orbital flight yet. Walter Schirra orbited the Earth six times in his Sigma 7 in what was a technical evaluation of the spacecraft's durability.

## 15 May 1963
## Mercury-Atlas 9: Faith 7

In Project Mercury's final crewed space mission, Gordon Cooper completed 22 Earth orbits in his Faith 7 space capsule. This would be the very last solo orbital mission.

# Project Gemini 1961-66

This second US human spaceflight programme set out to develop space travel techniques ahead of the Apollo mission to land humans on the Moon. The mission focused on human endurance in space, including 'extravehicular activities' without tiring, and perfecting rendezvous and docking with a second spacecraft.

### 23 Mar 1965
### Gemini 3: Molly Brown

The first of Project Gemini's missions saw Gus Grissom and John Young fly three low Earth orbits. As well as being the first time that two US astronauts had flown together, it was also the first time that humans had tested spacecraft manoeuverability, which they achieved through firing thrusters to change the shape and size of their orbit.

### 3 Jun 1965
### Gemini 4

The second Gemini mission entailed the first spacewalk with Ed White tethered to the spacecraft and floating freely, while his command pilot James McDivitt oversaw the first attempt to rendezvous with the Titan II, albeit unsuccessfully.

## 21 Aug 1965
## Gemini 5

This mission saw Gordon Cooper and Charles Conrad Jr. breaking the world record for the amount of time spent in space by any human. Their duration hit seven days, 22 hours and 55 mins – a record that would have been a day longer had Hurricane Betsy not cut it short.

## 15 Dec 1965
## Gemini 6A

Wally Schirra and Thomas Stafford flew Gemini SC6 to become the first astronauts to rendezvous with another spacecraft – Gemini 7. Although not equipped to dock, they did come within one foot.

## 4 Dec 1965
## Gemini 7

Frank Borman and James Lovell spent almost 14 days in space, making 206 orbits, and partaking as a passive target in the first rendezvous mission with sister spacecraft Gemini 6.

## 16 Mar 1966
## Gemini 8

This was the first time that two spacecraft docked in orbit. Unfortunately, it was also the first time that a US spacecraft had a critical system failure and Neil Armstrong and David Scott were forced to abort the mission in order to return safely to Earth.

## 3 Jun 1966
## Gemini 9A

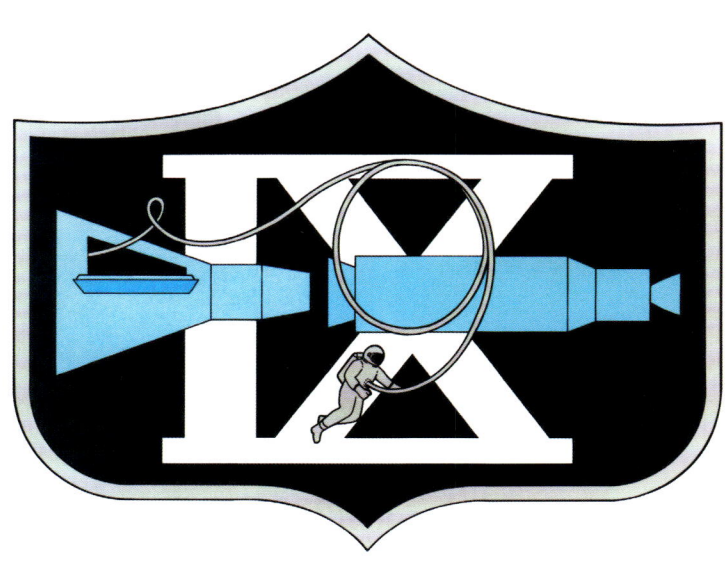

Both mission objectives were unsuccessful. After a failed attempt to dock with an uncrewed spacecraft, Eugene Cernan spent two hours in a self-contained rocket pack trying to 'spacewalk'. Due to fatigue, overheating and cardiac stress, Cernan was unable to complete the task and he and fellow astronaut Thomas Stafford aborted the mission.

## 18 Jul 1966
## Gemini 10

Crewed by John Young and Michael Collins, this mission set out to achieve rendezvous and docking with a second uncrewed spacecraft. Tethered to a 50-foot (15-metre) line, Collins became the first astronaut to spacewalk to another spacecraft in orbit.

## 12 Sep 1966
## Gemini 11

The objective for this mission was to rendezvous and dock on the Agena Target Vehicle immediately upon entry into orbit, which Charles Conrad and Richard Gordon duly achieved. The crew also completed an orbit of the furthest distance from Earth.

## 11 Nov 1966
## Gemini 12

The final Project Gemini mission concluded with a total of five hours and 30 minutes of extravehicular activity by an astronaut outside of the spacecraft in preparation for the next project to land man on the Moon. The crew for this mission was James Lovell and Buzz Aldrin.

# Project Apollo 1961-72

The Apollo programme was conceived to prepare US astronauts to be the first men to land on the Moon. This objective would be achieved on the sixth crewed spaceflight with mission Apollo 11 on 20 July 1969.

## 21 Feb 1966 — Apollo 1

The first crewed mission of Project Apollo was due to launch on 21 February 1967, but tragically, during a rehearsal of the launch, a cabin fire killed all three crew members, Virgil Grissom, Edward White and Roger Chaffee, as well as destroying the command module.

## 11 Oct 1968 — Apollo 7

This mission set out to test the command and service modules in low Earth orbit as a first step towards achieving the goal of landing on the Moon. Walter Schirra, Donn Eisele and Walter Cunningham completed the mission, which lasted 11 days, but it would be the last journey into space for all three men.

## 21 Dec 1968 — Apollo 8

Frank Borman, William Anders and James Lovell became the first humans to see the far side of the Moon. They orbited the Moon 10 times before returning safely to Earth.

## 3 Mar 1969 — Apollo 9

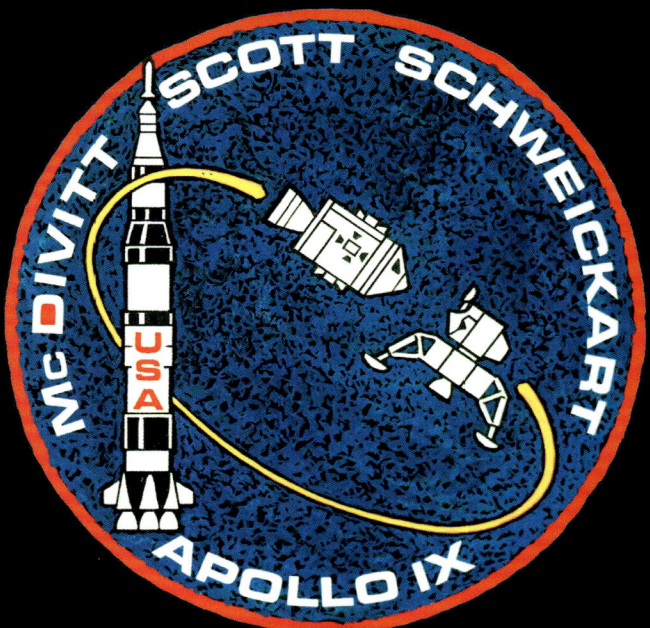

Flying in low orbit, James McDivitt, David Scott and Russell Schweickart set out to fly the complete Apollo spacecraft – the command and service module (CSM) with the Lunar Module (LM). They succeeded in flying the LM independently before rendezvousing and docking with the CSM again.

**18 May 1969
Apollo 10**

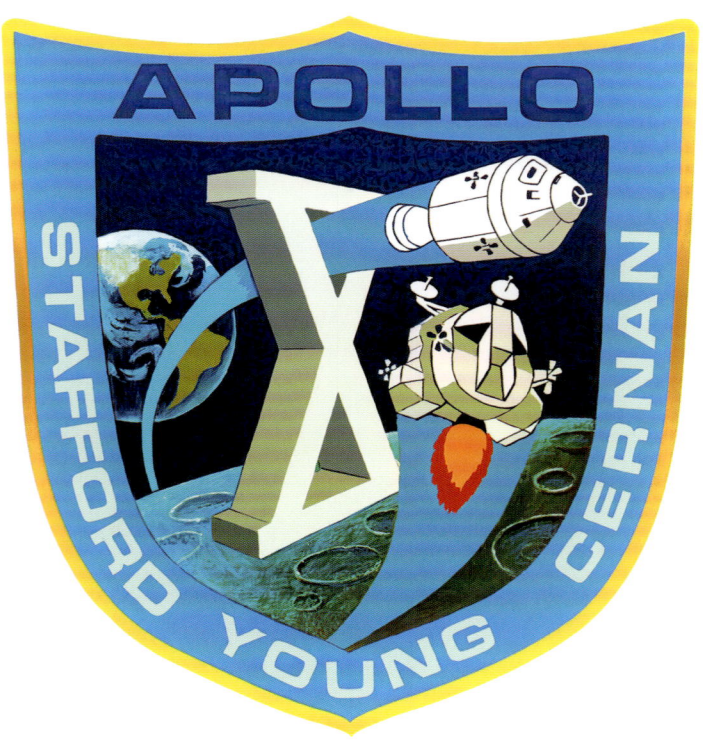

Essentially a dress rehearsal for the first Moon landing, Thomas Stafford, John Young and Gene Cernan were tasked with testing all spacecraft components and procedures.

**16 Jul 1969
Apollo 11**

The first scheduled lunar landing mission saw Neil Armstrong become the first human to land on the Moon, followed by Buzz Aldrin 19 minutes later. The third member of their crew, Michael Collins, flew the CSM in lunar orbit while Armstrong and Aldrin spent 21 hours and 36 minutes exploring the Moon's surface.

**14 Nov 1969
Apollo 12**

Following the success of Apollo 11, Charles Conrad, Richard Gordon and Alan Bean flew the second mission to land on the Moon. Conrad and Bean were on the lunar surface engaged in various activities for a total of one day and seven hours.

**11 Apr 1969
Apollo 13**

Setting out as the third crewed spacecraft to land on the Moon, the Apollo 13 mission was forced to abort following the rupture of an oxygen tank which rendered the electrical and life-support systems inoperative. The crew, James Lovell, John Swigert and Fred Haise, transferred to the LM and mission controllers worked to bring them back to Earth alive.

## 31 Jan 1971
## Apollo 14

Apollo 14 was the last of the Moon landing missions that targeted specific sites of scientific interest. On this mission, Alan Shepard, Stuart Roosa and Edgar Mitchell were the first humans to land in the lunar highlands.

## 26 Jul 1971
## Apollo 15

In this ninth crewed Apollo mission, Alfred Worden orbited the Moon collecting data, while David Scott and James Irwin explored the lunar surface in the Lunar Roving Vehicle, enabling them to travel further than previously possible and allowing them to stay for a total of 18 and a half hours on the Moon's surface.

## 16 Apr 1971
## Apollo 16

Spending 71 hours on the Moon, John Young and Charles Duke drove a Lunar Roving Vehicle for 16.6 miles (26.7 km) across the lunar surface, collecting samples to take back to Earth, while the third member of their crew, Thomas Mattingly, orbited the Moon, taking photos and retrieving scientific data.

## 7 Dec 1972
## Apollo 17

The final Apollo mission included a professional geologist, Harrison Schmitt, who joined Commander Gene Cernan in retrieving material from the lunar highland area of Taurus-Littrow, where it was thought that volcanic activity might have taken place. CSM Pilot Ronald Evans orbited above.

# Skylab 1973

The mighty Skylab was America's first orbital space station. The purpose was to test human responses to prolonged spaceflight, along with observations of various celestial bodies.

## 14 May 1973
## Skylab

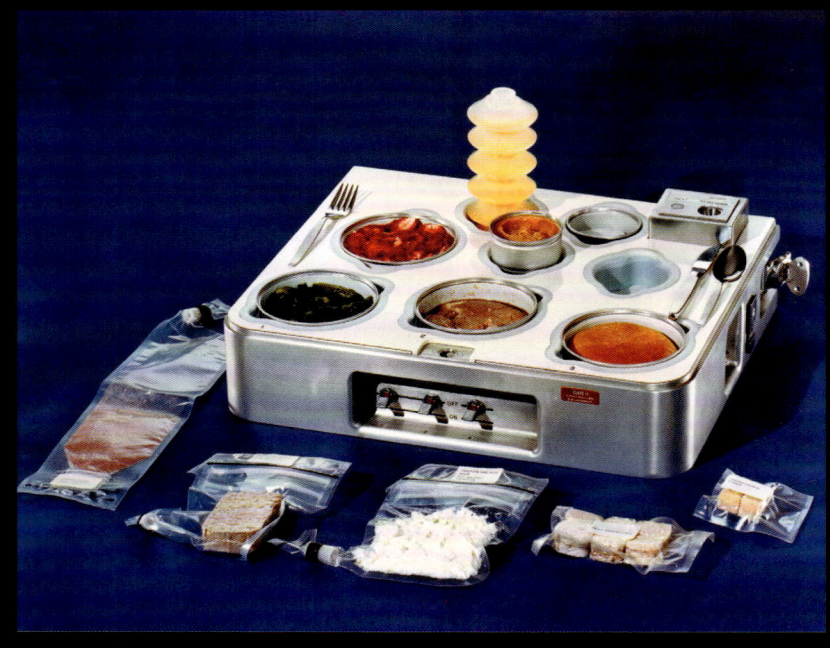

The first Skylab mission was uncrewed. Thereafter, there was some confusion with numerical identification which was changed at the last minute. As a result, mission patches representing crewed missions 2, 3 and 4 were designed as missions 1, 2 and 3.

## 25 May 1973
## Skylab 2

The first crewed mission to Skylab saw Pete Conrad, Joseph Kerwin and Paul Weitz take the record for the longest spaceflight at 28 days. Kerwin was chosen for his skills as a medical doctor with a view of getting a deeper understanding of the effect that spaceflight has on the human body.

## 28 Jul 1973
## Skylab 3

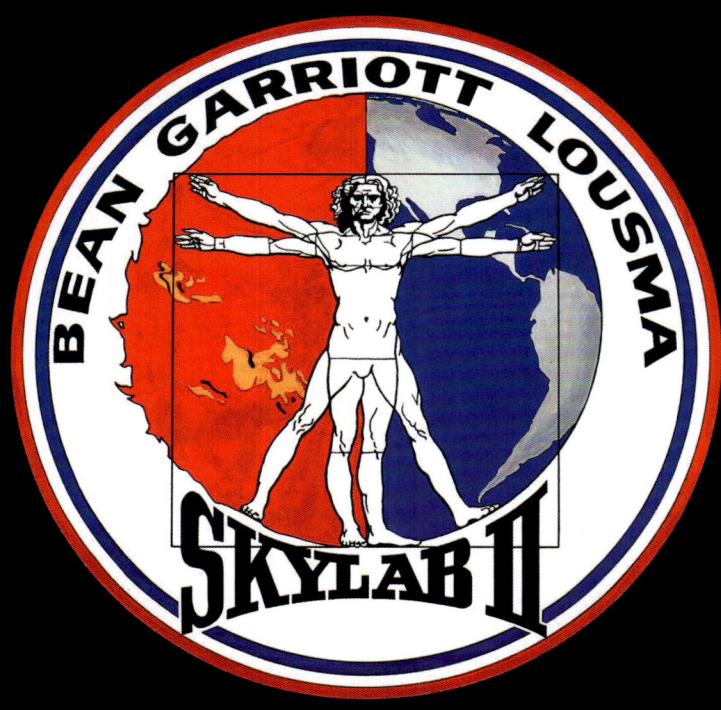

The second crewed mission to Skylab entailed Alan Bean, Owen Garriott and Jack Lousma performing a variety of tasks, including scientific experiments, solar observations and medical projects, all of which amounted to over 1,000 hours of astronaut activity.

## 16 Nov 1973
## Skylab 4

This final crewed mission to Skylab lasted nearly three months. Astronauts Gerald Carr, Edward Gibson and William Pogue continued with the activities undertaken by the crew of the previous mission but were also tasked with observing the comet Kohoutek.

# Apollo-Soyuz 1975

East met West in this historical show of peace and cooperation between rival superpowers.

### 15 Jul 1975
### Apollo-Soyuz Test Project

The Apollo-Soyuz project was the first international crewed mission into space. Thomas Stafford, Vance Brand and Keke Slayton from the USA launched in an Apollo spacecraft to dock with the Soviet Soyuz capsule crewed by Alexei Leonov and Valery Kubasov. Essentially, this mission was a political act of peace amid Cold War tensions.

# Chapter 2:
# Space Shuttles and Their Missions

*Columbia, Challenger, Discovery, Atlantis, Endeavour*

| | |
|---|---|
| **12 April 1981 – STS-1** | The first flight of NASA's Columbia space shuttle. This was the administration's first space shuttle flight and the first crewed American spaceflight since the Apollo-Soyuz linkup. Coincidentally, this date also marked the 20th anniversary of the first-ever human spaceflight, which was undertaken by Yuri Gagarin for the former Soviet Union. |
| **18 June 1983 – STS-6** | As a member of the Challenger shuttle crew, astronaut Sally Ride became the first American woman in space. |
| **30 August 1983 – STS-8** | Astronaut Guion Bluford Jr. became the first African American in space. |
| **28 January 1986 – STS-51L** | An incident that shocked America. Just 73 seconds after launch, the Challenger shuttle exploded, killing all seven astronauts onboard. The icy conditions were found to be partly responsible. The space shuttle programme was immediately suspended. |
| **29 September 1988 – STS-26** | The space shuttle programme returned when the Discovery (first launched in '84) retook to the skies, with more than a million people gathered on beaches nearby to watch the launch. |
| **4 May 1989 – STS-30** | Named after a 16th-century Portuguese explorer, the Magellan probe was released into outer space by the Atlantis shuttle and sent to map the surface of Venus. |
| **18 October 1989 – STS-34** | Atlantis deployed the Galileo probe. Bound for Jupiter, the Galileo was the first craft to orbit an outer planet and the first to visit an asteroid. |
| **21 June 1993 – STS-57** | The Endeavour shuttle (first launched just over a year earlier) carried the Spacehab into orbit for the first time. The commercially developed onboard lab doubled the available workspace for the crew. |
| **23 July 1999 – STS-93** | Having already been the first woman to pilot a space shuttle (the Discovery in 1995), Eileen Collins became the first woman to command a space shuttle, taking charge of the Columbia on a five-day mission to deploy the Chandra X-Ray Observatory. This was also the shortest scheduled shuttle mission since 1990. |
| **1 February 2003** | The Columbia was lost, disintegrating after re-entry into the Earth's atmosphere. All seven crewmembers were killed. Damaged panels on the underside of the left wing caused a catastrophic failure. The Columbia was just 15 minutes from its scheduled landing at the Kennedy Space Center. |
| **21 July 2011** | The end of the final space shuttle mission as the programme was brought to a close. |

# Space Shuttle Columbia 1981–2003

Space Shuttle Columbia was the first of five space shuttle orbiters – reusable spaceplane components that attach to the shuttle. Columbia was named after both the first American ship to circumnavigate the globe and 'Miss Columbia', the national personification of the USA. Columbia flew 28 missions over a period of 22 years, with a total of 4,000 orbits around the Earth over 300 days in space.

### 12 Apr 1981
### STS-1: Columbia

The first orbital shuttle flight was crewed by John Young and Robert Crippen and was essentially a maiden voyage to ensure that a shuttle orbiter could make it into orbit and return to Earth safely with both vehicle and crew intact. All objectives were achieved and the spaceworthiness of Columbia confirmed.

### 12 Nov 1981
### STS-2: Columbia

Crewed by Joe Engle and Richard Truly, Columbia's second mission marked the first time that a crewed spacecraft had returned to space. A multitude of experiments were carried out on this mission, testing out the shuttle's robotic arm, measuring air pollution from satellites and shutting off and restarting the shuttle's engines.

### 22 Mar 1982
### STS-3: Columbia

Jack Lousma and Gordon Fullerton flew Columbia's third mission, which, as well as carrying out a great many scientific experiments, was largely aimed at testing the endurance of the orbiter itself.

### 27 Jun 1982
### STS-4: Columbia

Crewed by Ken Mattingly and Henry Hartsfield, this was Columbia's final test flight before the orbiter was officially declared fully operational. This meant that ejection seats were deactivated and the astronauts no longer had to wear pressure suits.

## 11 Nov 1982
## STS-5: Columbia

Astronauts Vance Brand, Robert Overmyer, Joseph Allen and William Lenoir were given the objective of deploying communications satellites into orbit on what would be Columbia's first 'officially operational' mission. The five-pointed star of the mission badge is representative of this being Columbia's fifth mission in total.

## 28 Nov 1983
## STS-9: Columbia

Columbia's sixth mission had the largest crew yet, consisting of John Young, Brewster Shaw, Owen Garriott, Robert Parker, Byron Lichtenberg and Ulf Merbold (the first West German to go up in space). This 10-day mission carried the reusable laboratory, known as Spacelab, into orbit.

## 12 Jan 1986
## STS-61C: Columbia

Following four failed launches, Columbia finally got this satellite-deployment mission in motion. Alongside several other scientific experiments was a test on an artist's paintings. Ellery Kurtz had his work taken up into space to determine the effects that the space environment would have on the materials.

## 8 Aug 1989
## STS-28: Columbia

Astronauts Brewster Shaw Jr., Richard Richards, Mark Brown, James Adamson and David Leestma took the Columbia on its fourth US Department of Defense flight.

## 9 Jan 1990
## STS-32: Columbia

After almost a month of delays due to various issues with the launchpad and then poor weather conditions, Daniel Brandenstein, James Wetherbee, Marsha Ivins, David Low and Bonnie Dunbar set off to deploy a defense communications satellite and retrieve NASA's Long Duration Exposure Facility (LDEF). The crew also tested the circadian rhythms of a test-tube fungus to see if they changed in outerspace.

## 2 Dec 1990
## STS-35: Columbia

Initially delayed by a hydrogen leak, the Columbia's tenth spaceflight was crewed by Vance Brand, Guy Gardner, John Lounge, Robert Parker, Jeffrey Hoffman, Ronald Parise and Samuel Durrance. Lasting almost nine days, the mission objectives centred around the ultraviolet and x-ray telescopes in the onboard ASTRO-1 observatory.

### 5 Jun 1991
### STS-40: Columbia

Delayed by numerous gas leaks and computer failures, this was the fifth dedicated Spacelab mission and the first devoted entirely to life sciences. The crew included astronauts Bryan O'Connor, Sidney Gutierrez, Drew Gaffney, Millie Hughes-Fulford, Rhea Seddon, James Bagian and Tamara Jernigan, along with 30 rodents and thousands of miniscule jellyfish.

### 25 Jun 1992
### STS-50: Columbia

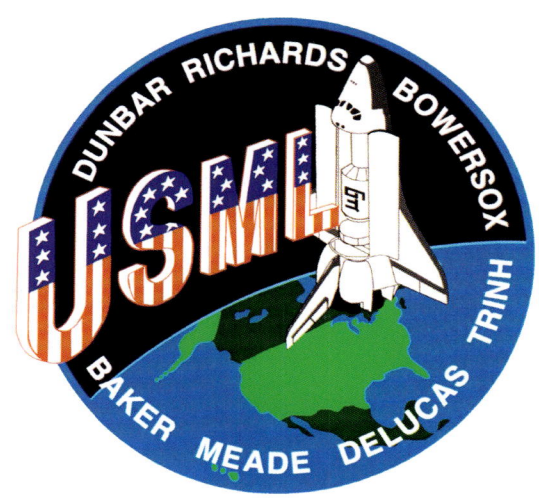

The crew of Richard Richards, Kenneth Bowersox, Bonnie Dunbar, Ellen Baker, Carl Meade, Lawrence DeLucas and Eugene Trinh spent 13 days and 19 hours in space, conducting multi-disciplinary microgravity experiments using the United States Microgravity Laboratory-1 (USML-1).

### 22 Oct 1992
### STS-52: Columbia

James Wetherbee, Michael Baker, Charles Veach, William Shepherd, Tamara Jernigan and Steven MacLean crewed this mission to deploy the Laser Geodynamic Satellite 2 (LAGEOS-2) and conduct further microgravity experiments.

### 26 Apr 1993
### STS-55: Columbia

This was a German-managed mission, featuring Steven Nagel, Terence Henricks, Jerry Ross, Charles Precourt, Bernard Harris, Ulrich Walter and Hans Schlegel, who carried out multiple experiments, including one in which a floating object was caught by a robot arm that was remotely controlled from Earth.

### 18 Oct 1993
### STS-58: Columbia

The scheduled launch on 14 October was 'scrubbed' (cancelled) due to a computer failure and the new 18 October launch was delayed by an aircraft in the launch zone. But once off the ground, John Blaha, Richard Searfoss, Rhea Seddon, William McArthur, David Wolf, Shannon Lucid and Martin Fettman proceeded to undertake the Columbia's second dedicated life-sciences mission with the help of 48 rats.

### 4 Mar 1994
### STS-62: Columbia

This near-14-day mission featured the USMP-2 (another microgravity lab) and astronauts John Casper, Andrew Allen, Pierre Thuot, Charles Gemar and Marsha Ivins.

## 8 Jul 1994
### STS-65: Columbia

Chiaki Naito-Mukai became the first Japanese woman in space and set a new record for the longest spaceflight by a woman (14 days, 17 hours and 55 minutes). The other crew members were Robert Cabana, James Halsell, Richard Hieb, Carl Walz, Leroy Chiao and Donal Thomas.

## 20 Oct 1995
### STS-73: Columbia

This mission tied with STS 61-C (Jan 1986) for most launch scrubs (six). Astronauts Kenneth Bowersox, Kent Rominger, Kathryn Thornton, Catherine Coleman, Michael Lopez-Alegria, Fred Leslie and Albert Sacco performed experiments in the fields of biotech, fluid physics and combustion science, among others. During their near-16 days in orbit, the crew managed to grow five small potatoes from tubers.

## 22 Feb 1996
### STS-75: Columbia

The 75th shuttle mission saw the deployment of an Italian-US Tethered Satellite System, as well as further microgravity experiments. The crew included Andrew Allen, Scott Horowitz, Franklin Chang-Díaz, Maurizio Cheli, Umberto Guidoni, Jeffrey Hoffman and Claude Nicollier.

## 20 Jun 1996
### STS-78: Columbia

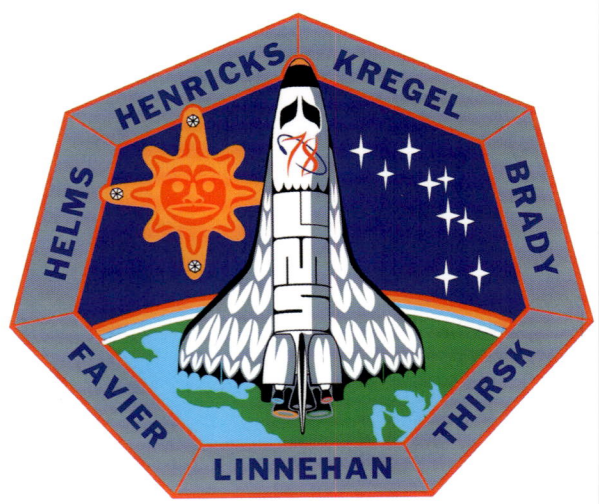

NASA, ESA, CSA, and the Italian Space Agency came together to create the combined Life and Microgravity Spacelab (LMS) for this mission. It was the first mission to provide video images from the flight deck. The crew of Terrence Henricks, Kevin Kregel, Jean-Jacques Favier, Richard Linnehan, Susan Helms, Charles Brady Jr. and Robert Thirsk tested several processes that would later be used aboard the ISS.

## 19 Nov 1996
### STS-80: Columbia

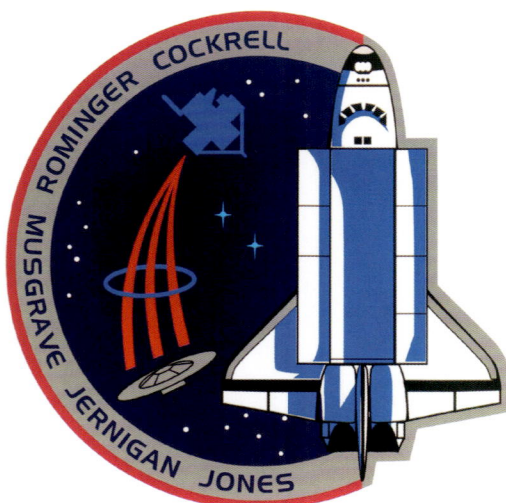

Kenneth Cockrell, Kent Rominger, Tamara Jernigan, Story Musgrave and Thomas Jones successfully deployed, operated and retrieved two free-flying research spacecrafts.

## 4 Apr 1997
### STS-83: Columbia

This mission was cut short after just three days because of a fuel cell malfunction. The crew - James Halsell, Susan Still, Janice Voss, Donald Thomas, Michael Gernhardt, Roger Crouch and Greg Linteris - did manage to conduct a handful of experiments, including a study into the properties of stabilised flame balls.

### 1 Jul 1997
### STS-94: Columbia

The first ever reflight featuring the same crew, payloads and vehicles as a previous mission, the rerun of STS-83 was a success. Experiments included the growth of protein crystals, accelerated by conditions in outer space, with the intention of advancing research into cures for diabetes, AIDS and cancer.

### 19 Nov 1997
### STS-87: Columbia

The crew of Kevin Kregel, Steven Lindsey, Kalpana Chawla, Leonid Kadenyuk, Winston Scott and Takao Doi ran into trouble when deploying the SPARTAN-201-04 free-flying spacecraft. The craft failed to manoeuvre correctly and span out of control when they tried to regrapple it. Scott and Doi then performed a seven-hour, 43-minute spacewalk to retrieve the errant craft by hand, with Doi becoming the first Japanese person to walk in space.

### 17 Apr 1998
### STS-90: Columbia

Richard Searfoss, Scott Altman, Richard Linnehan, Dafydd Rhys Williams, Kathryn Hire, Jay Buckey Jr. and James Pawelczyk took off on 17 April with a menagerie of rats, mice, crickets, snails and fish. NASA was aided by CSA, CNES, DARA and ESA, with the mission objectives focused on advancing our understanding of neuroscience.

### 23 Jul 1999
### STS-93: Columbia

The first shuttle mission commanded by a woman was delayed for around three days after a hair-raising false start. The initial launch was scrubbed at the T-7 mark when a potential hydrogen leak was detected in the aft engine compartment. The launch sequencer was manually cutoff with just half a second to spare. Commander Eileen Collins was joined by Jeffrey Ashby, Steven Hawley, Catherine Coleman and Michel Tognini.

### 1 Mar 2002
### STS-109: Columbia

A mission to refurbish the Hubble telescope during a series of spacewalks. The telescope was successfully pulled into the payload bay and kitted out with updated equipment by a crew of Scott Altman, Duane Carey, John Grunsfeld, Nancy Currie, James Newman, Richard Linnehan and Michael Massimino.

### 16 Jan 2003
### STS-107: Columbia

The final flight of Columbia. This 16-day mission featured an extraordinary array of studies into water recycling, extraterrestrial crop yields, viruses, bone formation, the spaceflight responses of spiders, silkworms, inorganic crystals, fish, bees and ants, the effect of microgravity on bubbles and droplets, cardiopulminary changes in astronauts and much more. But tragically, the left wing had been damaged during the launch sequence and, upon re-entry into the Earth's atmosphere, the Columbia began to break up. The crew of Rick Husband, William McCool, Kalpana Chawla, David Brown, Laurel Clark, Michael Anderson and Ilan Ramon were lost in the skies over Texas.

# Space Shuttle Challenger 1983-86

Challenger was only ever intended as a test vehicle, but after an upgrade made it mission-ready, the orbiter rose to become leader of the shuttle fleet. During three years of service, Challenger completed 9 flights, carrying 60 different astronauts and spending almost 70 days in space before its tenth mission ended in disaster.

### 4 Apr 1983
### STS-6: Challenger

The inaugural flight of Challenger was postponed due to a hydrogen leak in the main engine. The primary objective was the deployment of the first Tracking and Data Relay Satellite (TDRS-1). The crew were Paul Weitz, Karol Bobko, Donald Peterson and Story Musgrave.

### 18 Jun 1983
### STS-7: Challenger

Mission specialist Sally Ride became the first US woman to fly in space. The other crew members were Robert Crippen, Frederick Hauck, John Fabian and Norman Thagard. The mission included the deployment of two coms satellites and an experiment studying the effects of zero gravity on the social behaviours of an ant colony.

### 30 Aug 1983
### STS-8: Challenger

Guion Bluford Jr. became the first African American to fly in space. Richard Truly, Daniel Brandenstein, Dale Gardner and William Thornton made up the rest of the crew and the mission highlights included a test of the flight deck's response to extreme cold, whereby the shuttle's nose was aimed away from the sun for 14 hours.

### 3 Feb 1984
### STS-41B: Challenger

Vance Brand, Robert Gibson, Bruce McCandless II, Ronald McNair and Robert Stewart took the Challenger space shuttle up for a little over a week, during which time McCandless and Stewart carried out the very first untethered spacewalks.

## 6 Apr 1984
## STS-41C: Challenger

The first direct ascent trajectory for a space shuttle. The crew of Robert Crippen, Francis Scobee, George (Pinky) Nelson, James van Hoften and Terry Hart carried out repairs to the Solar Max satellite while in orbit.

## 5 Oct 1984
## STS-41G: Challenger

The first space flight to include two women proved that it's possible to refuel satellites in orbit. Robert Crippen, Jon McBride, David Leestma, Sally Ride, Kathryn Sullivan, Paul Scully-Power and Marc Garneau also deployed the Earth Radiation Budget Satellite, while Kathryn Sullivan became the first American woman to walk in space.

## 29 Apr 1985
## STS-51B: Challenger

With Spacelab-3 onboard, as well as two monkeys and 24 rodents, Robert Overmyer, Frederick Gregory, Don Lind, Norman Thagard, William Thornton, Lodewijk van den Berg and Taylor Wang carried experiments in the disciplines of materials sciences, life sciences, fluid mechanics, atmospheric physics and astronomy.

## 29 Jul 1985
## STS-51F: Challenger

Spacelab-2 carried an igloo into space as well as multiple scientific instruments. The crew members were Gordon Fullerton, Roy Bridges, Story Musgrave, Karl Henize, Anthony England, Loren Acton and John-David Bartoe.

## 30 Oct 1985
## STS-61A: Challenger

Henry Hartsfield Jr., Steven Nagel, James Buchli, Guion Bluford Jr., Bonnie Dunbar, Reinhard Furrer, Ernst Messerschmid and Wubbo Ockels crewed this mission, which was devoted to the West German Spacelab D-1.

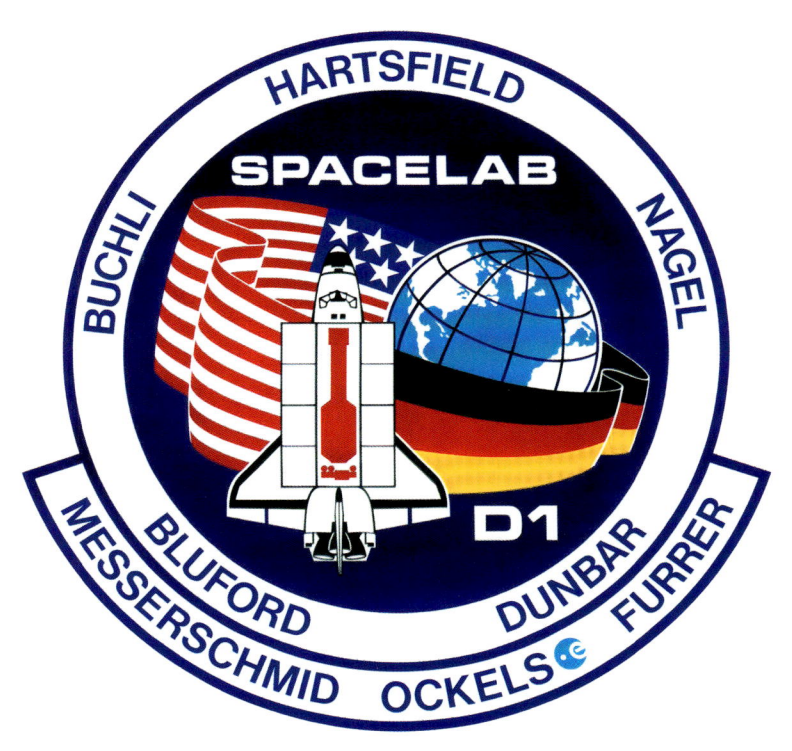

## 28 Jan 1986
## STS-51L: Challenger

Francis Scobee, Michael Smith, Judith Resnik, Ellison Onizuka, Ronald McNair, Greg Jarvis and the first teacher in space Christa McAuliffe were scheduled to perform the 25th mission of the space shuttle programme. The objective was to deploy a TDRS satellite into orbit, to deploy and later retrieve a satellite designed to track the journey of Halley's Comet, and for McAuliffe to broadcast two lessons from outer space for school children across America. But 73 seconds after launch, Mission Control lost all coms and readings from the Challenger and watched in horror as a fireball lit up the TV screens. None of the crew members survived.

# Space Shuttle Discovery 1984-2011

Discovery was NASA's third space shuttle. Making its first public appearance during a special rollout ceremony on 16 October 1983, the orbiter outlived its predecessors and continued to ferry astronauts to and from outer space until the shuttle programme wrapped up in 2011.

### 30 Aug 1984
### STS-41D: Discovery

Discovery's first flight with crew members Henry Hartsfield Jr., Michael Coats, Judith Resnik, Richard Mullane, Steven Hawley and Charles Walker. The payload included the 102-foot-tall OAST-1 solar wing, which carried an array of different solar cells.

### 8 Nov 1984
### STS-51A: Discovery

Frederick Hauck, David Walker, Anna Fisher, Dale Gardner and Joseph Allen crewed this mission, during which Allen and Gardner used jetpacks to retrieve two malfunctioning satellites.

### 24 Jan 1985
### STS-51C: Discovery

The crew were Thomas Mattingly II, Loren Shriver, James Buchli, Ellison Onizuka and Gary Payton. Originally penned for Challenger, this mission was most notable for a handful of minor technical issues.

### 12 Apr 1985
### STS-51D: Discovery

Karol Bobko, Donald Williams, Rhea Seddon, David Griggs, Jeffrey Hoffman and Charles Walker were joined onboard by US Senator E.J. Garn and an informal study was conducted into the behaviour of certain toys in space. Brake damage and a blown tire during landing led to changes in future landing protocols and equipment.

**17 Jun 1985**
**STS-51G: Discovery**

The 18th flight of the shuttle programme. The crew included Sultan Salman Al-Saud, along with Daniel Brandenstein, John Creighton, Shannon Lucid, Steven Nagel, John Fabian and Patrick Baudry. The primary objective was the deployment of ARABSAT-A for an Arabic communications network, along with two other coms satellites.

**27 Aug 1985**
**STS-51I: Discovery**

The crew were Joe Engle, Richard Covey, James van Hoften, John Lounge and William Fisher. The mission focused on the deployment, retrieval, repair and rerelease of various satellites.

**29 Sep 1988**
**STS-26: Discovery**

The first time Discovery had flown since the Challenger disaster. Frederick Hauck, Richard Covey, David Hilmers, George (Pinky) Nelson and John Lounge crewed the four-day mission and suffered slightly adverse conditions onboard when a fluid cooling system iced up, causing the cabin temperature to rise to almost 30ºC.

**13 Mar 1989**
**STS-29: Discovery**

Michael Coats, John Blaha, James Bagian, Robert Springer and James Buchli crewed this mission to deploy another Tracking and Data Relay Satellite (TDRS).

**22 Nov 1989**
**STS-33: Discovery**

The third night launch and the first since the shuttle programme's restart in the wake of the Challenger disaster. Crewed by Frederick Gregory, John Blaha, Kathryn Thornton, Story Musgrave and Manley Carter Jr., it was also the fifth mission dedicated to the Department of Defense.

### 24 Apr 1990
### STS-31: Discovery

A momentous occasion in the history of space exploration. The primary payload was the Hubble Space Telescope, which was successfully deployed into orbit. Loren Shriver, Charles Bolden Jr., Steven Hawley, Bruce McCandless II and Kathryn Sullivan crewed this important mission.

### 6 Oct 1990
### STS-41: Discovery

Built by ESA, the Ulysses probe was deployed on this mission to explore the sun's polar regions. The crew members were Richard Richards, Robert Cabana, Bruce Melnick, Thomas Akers and William Shepherd.

### 28 Apr 1991
### STS-39: Discovery

A Department of Defense mission, carried out by Michael Coats, Blaine Hammond, Guion Bluford Jr., Gregory Harbaugh, Richard Hieb, Donald McMonagle and Charles Veach.

### 12 Sep 1991
### STS-48: Discovery

The main objective was the deployment of a research satellite. John Creighton, Kenneth Reightler Jr., Mark Brown, Charles Gemar and James Buchli were the crew members.

### 22 Jan 1992
### STS-42: Discovery

The crew of Ronald Grabe, Stephen Oswald, Norman Thagard, David Hilmers, William Readdy, Roberta Bondar and Ulf Merbold were divided into two teams to make round-the-clock observations on the human nervous system and its adaptation to low gravity. They also monitored shrimp eggs, lentil seedlings, fruit fly eggs, and bacteria. The mission was extended by one day after mission managers concluded the onboard consumables remained at a suitable level.

### 2 Dec 1992
### STS-53: Discovery

A classified DoD mission with crew members David Walker, Robert Cabana, Guion Bluford Jr., James Voss and Michael Clifford, launched from the Kennedy Space Center.

33

### 8 Apr 1993
### STS-56: Discovery

Kenneth Cameron, Stephen Oswald, Michael Foale, Kenneth Cockrell and Ellen Ochoa crewed this voyage with ATLAS-2 and SPARTAN-201 payloads.

### 12 Sep 1993
### STS-51: Discovery

Launch was delayed and scrubbed several times for various reasons, including the Perseid meteor shower caused by a passing comet. The crew members were Frank Culbertson, William Readdy, James Newman, Daniel Bursch and Carl Walz.

### 3 Feb 1994
### STS-60: Discovery

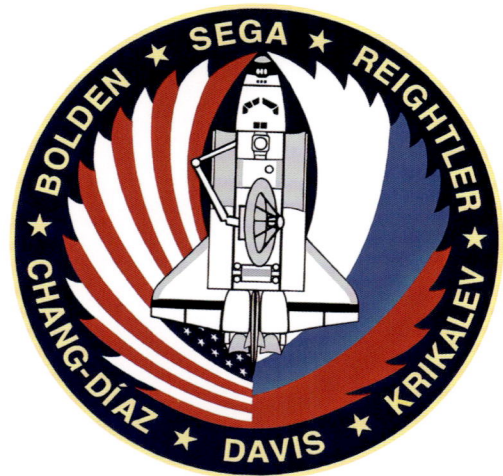

This was the first mission of the Shuttle-Mir Program - a joint US-Russian project. Russian cosmonaut Sergei Krikalev joined US crew members Charles Bolden Jr., Kenneth Reightler Jr., Franklin Chang-Díaz, Jan Davis and Ronald Sega to carry out in-flight radiological and medical tests.

### 9 Sep 1994
### STS-64: Discovery

Richard Richards, Blaine Hammond Jr., Jerry Linenger, Susan Helms, Carl Meade and Mark Lee conducted the Lidar In-space Technology Experiment (LITE), wherein laser pulses were used to study Earth's atmosphere, yielding images of cloud structures, storm systems, dust clouds, pollutants, forest burning and surface reflectance. Meade and Lee conducted the first untethered spacewalk by US astronauts since STS-41B, testing new gear for rescuing astronauts who become accidentally untethered.

### 3 Feb 1995
### STS-63: Discovery

First shuttle mission piloted by a woman, the first shuttle to approach and fly around the Russian space station Mir and the second time a Russian cosmonaut had flown aboard a US shuttle. James Wetherbee, Eileen Collins, Janice Voss, Bernard Harris Jr., Michael Foale and Vladimir Titov crewed the flight.

### 13 Jul 1995
### STS-70: Discovery

The schedule for this mission was altered numerous times, including after a Memorial Day holiday weekend when Northern Flicker Woodpeckers poked some 200 holes in the foam insulation of the external tank (the giant disposable rocket used for takeoff). Crew members were Terrence Henricks, Kevin Kregel, Nancy Jane Currie, Donald Thomas and Mary Ellen Weber.

### 11 Feb 1997
### STS-82: Discovery

Second mission to repair and upgrade the Hubble Space Telescope. Kenneth Bowersox, Scott Horowitz, Mark Lee, Steven Hawley, Gregory Harbaugh, Steven Smith and Joseph Tanner crewed the flight. Hawley had originally deployed the telescope in 1990 and was responsible for retrieving it. Work was carried out via a series of EVAs.

### 7 Aug 1997
### STS-85: Discovery

The fourth NASA-DARA collaborative mission with Curtis Brown, Kent Rominger, Robert Curbeam, Stephen Robinson, Jan Davis and Bjarni Tryggvason as the crew.

### 2 Jun 1998
### STS-91: Discovery

Discovery docked to the Mir for the first time and brought home astronaut Andy Thomas, who had been living on the Mir for 130 days and officially joined the Discovery crew upon boarding. The rest of the crew members were Charles Precourt, Dominic Gorie, Wendy Lawrence, Franklin Chang-Díaz, Janet Kavandi and Valeriy Ryumin.

### 29 Oct 1998
### STS-95: Discovery

John Glenn, aged 77, became the oldest person to fly in space, having previously set a record, all the way back in 1962, as the first human in Earth orbit. His crewmates this time were Curtis Brown, Steven Lindsey, Scott Parazynski, Stephen Robinson, Pedro Duque and Chiaki Mukai.

### 27 May 1999
### STS-96: Discovery

The second ISS assembly and logistics mission. An experienced crew of Kent Rominger, Rick Husband, Ellen Ochoa, Tamara Jernigan, Daniel Barry, Julie Payette and Valery Ivanovich Tokarev transferred clothing, sleeping bags, medical equipment, hardware, water and other essentials to the interior. Jernigan and Barry also attached a crane to the exterior during the second-longest spacewalk ever conducted up to that point.

### 19 Dec 1999
### STS-103: Discovery

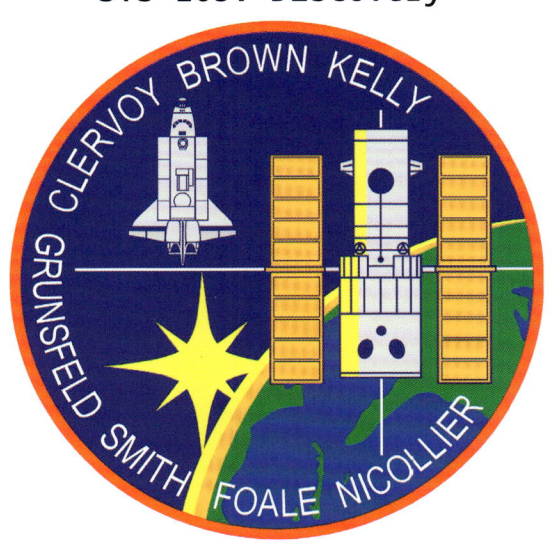

The third mission to service the Hubble Space Telescope. Curtis Brown, Scott Kelly, Steven Smith, Michael Foale, John Grunsfeld, Claude Nicollier and Jean-François Clervoy were the crew members.

### 11 Oct 2000
### STS-92: Discovery

Another mission to the ISS, involving four lengthy spacewalks, with a crew of Brian Duffy, Pamela Melroy, Koichi Wakata, Leroy Chiao, Peter Wisoff, Michael Lopez-Alegria and William McArthur.

### 8 Mar 2001
### STS-102: Discovery

A mission to rotate the ISS crew. Discovery was crewed by James Wetherbee, James Kelly, Andrew Thomas and Paul Richards. They transported Yury Usachev, James Voss and Susan Helms to the ISS and brought back William Shepherd, Yuri Gidzenko and Sergei Krikalev. The orbiter docked at the station for eight days while various modifications and installments were made before Shepherd formally handed command of the station over to Usachev, finalising the transfer.

### 10 Aug 2001
### STS-105: Discovery

The 11th ISS assembly mission also served as an opportunity to rotate the ISS crew. Scott Horowitz, Frederick Sturckow, Patrick Forrester and Daniel Barry crewed the Discovery. They went with Frank Culbertson Jr., Mikhail Tyurin and Vladimir Dezhurov and came back with Yury Usachev, James Voss and Susan Helms.

### 26 Jul 2005
### STS-114: Discovery

The first spaceflight after the loss of Columbia. It was dubbed the 'Return to Flight' mission and crewed by Eileen Collins, James Kelly, Wendy Lawrence, Stephen Robinson, Andrew Thomas, Charles Camarda and Soichi Noguchi.

### 4 Jul 2006
### STS-121: Discovery

Steven Lindsey, Mark Kelly, Piers Sellers, Michael Fossum, Lisa Nowak, Stephanie Wilson and Thomas Reiter carried out repairs to the ISS, as well as analysing safety improvements that were introduced for the post-Columbia 'Return to Flight' mission.

### 9 Dec 2006
### STS-116: Discovery

A mission to the ISS and the first night launch in more than four years, crewed by Mark Polansky, William Oefelein, Thomas Reiter, Joan Higginbotham, Robert Curbeam, Nicholas Patrick, Sunita Williams and Christer Fuglesang.

### 23 Oct 2007
### STS-120: Discovery

A near-perfect, on-time launch started this successful mission to the ISS, crewed by Pam Melroy, George Zamka, Scott Parazynski, Doug Wheelock, Stephanie Wilson and Paolo Nespoli, as well as Daniel Tani, who was dropped off at the station, and Clay Anderson, who was brought back.

### 31 May 2008
### STS-124: Discovery

Mark Kelly, Ken Ham, Karen Nyberg, Ron Garan, Mike Fossum, Akihiko Hoshide, Gregory Chamitoff and Garrett Reisman crewed this delivery of a robotic arm to the ISS.

### 15 Mar 2009
### STS-119: Discovery

An important mission to the ISS, installing solar arrays and upgrading the water recycling facility, allowing the station crew to double in size to six. President Obama called and spoke to the shuttle and station crew members, along with members of congress and a number of students. Lee Archambault, Toni Antonelli, Richard Arnold, Joseph Acaba, John Phillips and Steve Swanson crewed the Discovery, with Kiochi Wakata going up and Sandra Magnus coming down.

### 28 Aug 2009
### STS-128: Discovery

Mission to maintain the ISS, as well as carrying out some assembly work. The crew were Rick Sturckow, Kevin Ford, Jose Hernandez, John Olivas, Nicole Stott, Christer Fuglesang and Patrick Forrester.

### 5 Apr 2010
### STS-131: Discovery

Alan Poindexter, James Dutton, Rick Mastracchio, Stephanie Wilson, Dorothy Metcalf-Lindenburger, Naoko Yamazaki and Clayton Anderson crewed this 15-day flight to the ISS. Adding their numbers to the six crew members already aboard the station made a total of 13 people and marked the first time that four women had flown together in space.

### 24 Feb 2011
### STS-133: Discovery

The final flight of Discovery. An almost flawless mission, the shuttle touched down at the Kennedy Space Center after 27 years of service. The final crew members were Steve Lindsey, Eric Boe, Alvin Drew, Nicole Stott, Michael Barratt and Stephen Bowen, who replaced Tim Kopra a month before lift-off.

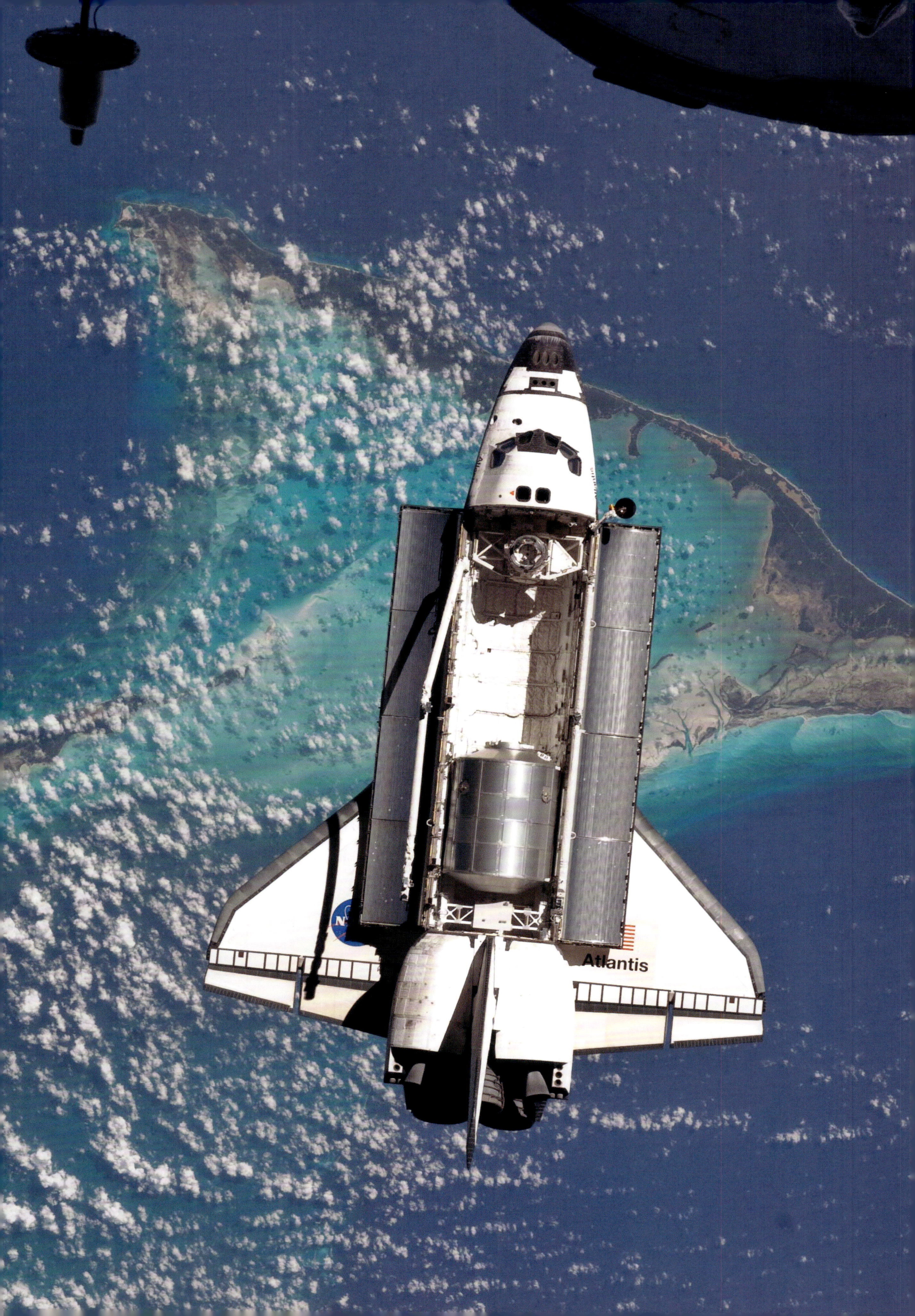
Atlantis

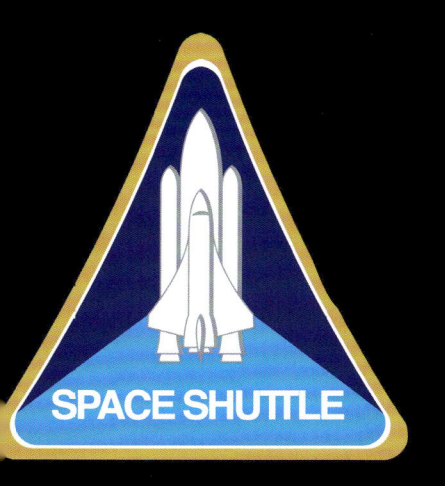

# Space Shuttle Atlantis 1984-2011

Another great survivor of the shuttle programme, Atlantis escaped a near-death experience on only its third mission and went on to be the very last space shuttle ever to fly. Safely touching down in Florida on the morning of 21 July 2011, the final Atlantis mission marked the conclusion of the space shuttle programme.

## 3 Oct 1985
### STS-51J: Atlantis

The first flight of the Atlantis was a classified DoD mission, crewed by Karol Bobko, Ronald Grabe, Robert Stewart, David Hilmers and William Pailes.

## 26 Nov 1985
### STS-61B: Atlantis

A mission to deploy three coms satellites with crew members Brewster Shaw, Bryan O'Connor, Mary Cleave, Sherwood Spring, Jerry Ross, Rodolpho Neri Vela and Charles Walker.

## 2 Dec 1988
### STS-27: Atlantis

Debris from the solid rocket booster struck Atlantis during takeoff, causing critical damage to around 700 heat-shield tiles. The classified nature of this DoD mission complicated in-flight efforts to assess the damage. It is believed that the crew of Robert Gibson, Guy Gardner, Jerry Ross, William Shepherd and Richard Mullane were extremely fortunate to survive re-entry and return home safely.

## 4 May 1989
### STS-30: Atlantis

Deployment of the Magellan spacecraft, which began its 15-month voyage to the planet Venus. Crew members were David Walker, Ronald Grabe, Norman Thagard, Mary Cleave and Mark Lee.

## 18 Oct 1989
## STS-34: Atlantis

Deployed from Atlantis, the Galileo spacecraft began its six-year journey to Jupiter. The mission was crewed by Donald Williams, Michael McCulley, Shannon Lucid, Franklin Chang-Díaz and Ellen Baker.

## 28 Feb 1990
## STS-36: Atlantis

Another classified DoD mission. Launch was briefly delayed while commander John Creighton recovered from a minor illness – the first time a manned spaceflight had been affected by illness to a crew member since Apollo 13. His crewmates were John Casper, Pierre Thuot, Richard Mullane and David Hilmers.

## 15 Nov 1990
## STS-38: Atlantis

The hydrogen leak discovered on Columbia during preparations for STS-35 set off a series of delays for STS-38. Precautionary tests on Atlantis showed that it too had a leaky hydrogen fuel tank. During repairs, the orbiter suffered damage from a falling platform beam and a passing hailstorm. The crew were Richard Covey, Frank Culbertson, Charles Gemar, Robert Springer and Carl Meade.

## 5 Apr 1991
## STS-37: Atlantis

This mission included the deployment of the Gamma Ray Observatory (GRO) and the first spacewalk since 1985. The crew were Steven Nagel, Kenneth Cameron, Jay Apt, Jerry Ross and Linda Godwin.

### 2 Aug 1991
### STS-43: Atlantis

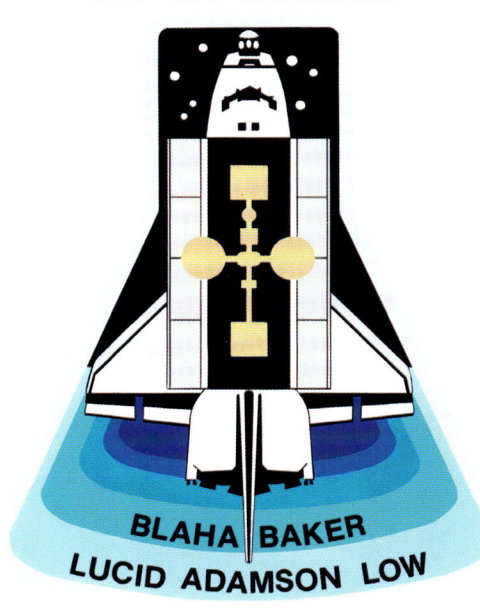

Mission to deploy satellites and perform various experiments. John Blaha, Michael Baker, Shannon Lucid, James Adamson and David Low were the crew.

### 24 Nov 1991
### STS-44: Atlantis

A DoD mission, including the deployment of a Defense Support Program (DSP) satellite. The crew members were Frederick Gregory, Terence Henricks, Story Musgrave, James Voss, Thomas Hennen and Mario Runco Jr.

### 24 Mar 1992
### STS-45: Atlantis

Charles Bolden Jr., Brian Duffy, Kathryn Sullivan, David Leestma, Michael Foale, Byron Lichtenberg and Dirk Frimout transported and operated the first Atmospheric Laboratory for Applications and Science (ATLAS-1). This was a non-deployable payload, featuring instruments from the USA, France, Germany, Belgium, Switzerland, the Netherlands and Japan.

### 31 Jul 1992
### STS-46: Atlantis

The deployment of ESA's Retrievable Carrier (EURECA) was the primary mission objective. A series of technical hitches delayed the EURECA from reaching its intended orbit, but the crew of Loren Shriver, Andrew Allen, Jeffrey Hoffman, Franklin Chiang-Díaz, Claude Nicollier, Marsha Ivins and Franco Malerba eventually succeeded on the sixth day.

### 3 Nov 1994
### STS-66: Atlantis

Atlantis underwent an extended service with Rockwell between October 1992 and May '94. This was the shuttle's first post-check-up launch and was crewed by Donald McMonagle, Curtis Brown, Ellen Ochoa, Scott Parazynski, Joseph Tanner and Jean-François Clervoy

## 27 Jun 1995
## STS-71: Atlantis

This mission represented the 100th human space launch. It was also the first time a space shuttle had docked with the Mir space station. Once docked, Atlantis and Mir represented the largest combined spacecraft ever in orbit. After a welcome ceremony staged on Mir, the station's crew was rotated – the first time this had been done in orbit. The shuttle crew members were Robert Gibson, Charles Precourt, Ellen Baker, Gregory Harbaugh and Bonnie Dunbar. They took Anatoly Solovyev and Nikolai Budarin up and brought Gennady Strekalov, Vladimir Dezhurov and Norman Thagard back.

## 12 Nov 1995
## STS-74: Atlantis

The second docking of a space shuttle to the Russian Mir station. Kenneth Cameron, James Halsell Jr., William McArthur Jr., Jerry Ross and Chris Hadfield crewed the flight.

## 22 Mar 1996
## STS-76: Atlantis

Veteran astronaut Shannon Lucid became the first woman to live aboard Mir and kicked off two years of continual US presence in space. Her Atlantis crewmates were Kevin Chilton, Rick Searfoss, Ron Sega, Michael Clifford and Linda Godwin.

## 16 Sep 1996
## STS-79: Atlantis

Looming threats from hurricanes Bertha and Fran delayed this mission to bring Shannon Lucid back from Mir. Her 188-day stint aboard the Russian station set a US spaceflight record and a world record for women. John Blaha travelled in the opposite direction, replacing Lucid on Mir. Their crewmates on Atlantis were William Readdy, Terrence Wilcutt, Jay Apt, Thomas Akers and Carl Walz.

## 12 Jan 1997
## STS-81: Atlantis

John Blaha returned to Earth and was replaced aboard Mir by Jerry Linenger. The rest of the Atlantis crew members were Michael Baker, Brent Jett, John Grunsfeld, Peter Wisoff and Marsha Ivins. They also brought back the first plants ever to complete a full life cycle in space – a crop of wheat planted as seeds by astronaut Shannon Lucid.

## 15 May 1997
## STS-84: Atlantis

This was the sixth time a shuttle had docked with Mir. Jerry Linenger was brought back to Earth, while Michael Foale took his place aboard the station – the fourth successive US member of the Mir crew. Charles Precourt, Eileen Collins, Edward Lu, Jean-François Clervoy, Elena Kondakova and Carlos Noriega made up the rest of the Atlantis crew.

## 25 Sep 1997
## STS-86: Atlantis

Atlantis headed back to Mir to collect Michael Foale and deliver his replacement, David Wolf. Foale had experienced the collision between a Progress (Russian) cargo craft and Mir's Spektr module, precipitating a number of issues. Part of Wolf's mission would be to initiate repairs. The rest of the Atlantis crew members were James Wetherbee, Mike Bloomfield, Wendy Lawrence, Vladimir titov, Scott Parazynski and Jean-Loup Chretien.

## 19 May 2000
## STS-101: Atlantis

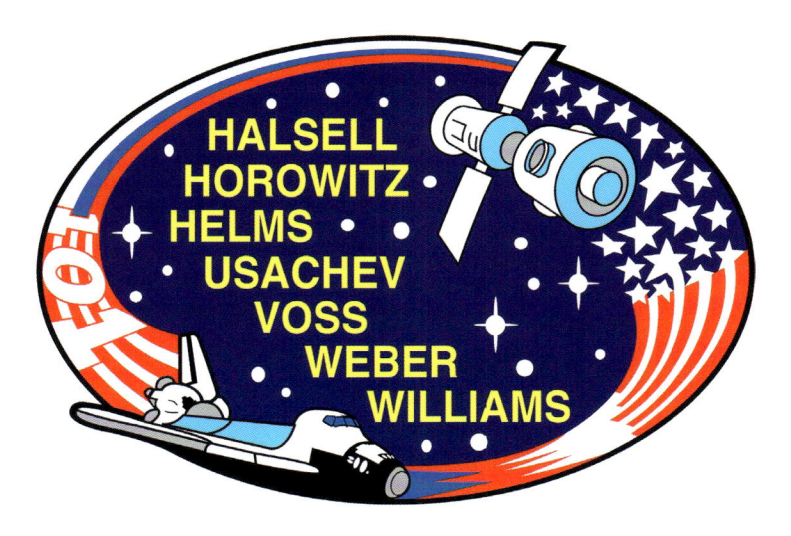

James Halsell Jr., Scott Horowitz, Mary Ellen Weber, Jeffrey Williams, Yury Usachev, James Voss and Susan Helms flew this mission to restock the ISS and carry out refurbishments to the Zarya and Unity modules.

## 8 Sep 2000
## STS-106: Atlantis

The Atlantis crew – Terrence Wilcutt, Scott Altman, Boris Morukov, Richard Mastracchio, Edward Lu, Daniel Burbank and Yuri Malechenko – delivered supplies and helped unload the contents of a Progress cargo vehicle onto the ISS.

## 7 Feb 2001
## STS-98: Atlantis

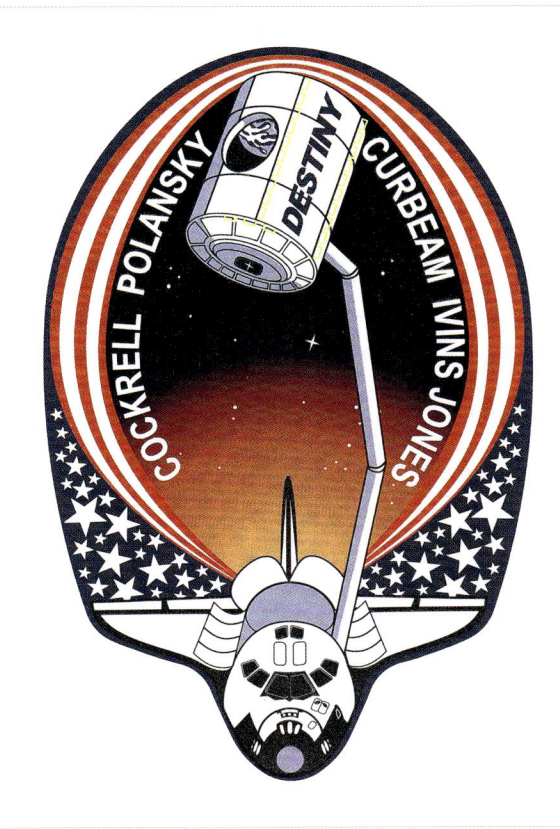

Successful mission to install the USA's Destiny laboratory module on the ISS. Kenneth Cockrell, Mark Polansky, Marsha Ivins, Robert Curbeam and Thomas Jones crewed the flight.

### 12 Jul 2001
### STS-104: Atlantis

Mission to the ISS with crew Steven Lindsey, Charles Hobaugh, Michael Gernhardt, Janet Kavandi and James Reilly.

### 8 Apr 2002
### STS-110: Atlantis

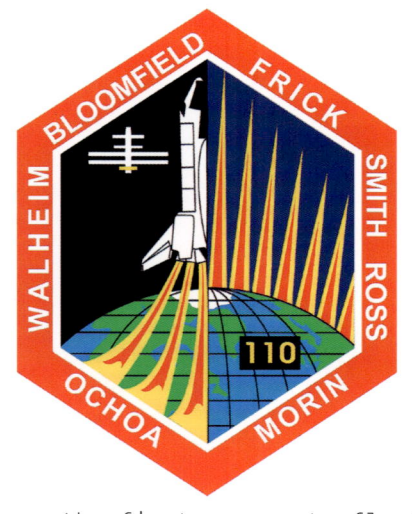

Jerry Ross became the first person to fly in space seven times, breaking his own record held jointly with other astronauts. His two spacewalks on this mission meant that only cosmonaut Anatoly Solovyev had spent more time walking in space. Ross's crewmates on this mission to the ISS were Michael Bloomfield, Stephen Frick, Steven Smith, Ellen Ochoa, Lee Morin and Rex Walheim.

### 7 Oct 2002
### STS-112: Atlantis

Mission to the ISS to install the Crew and Equipment Translation Aid (CETA), the first of two carts designed to run along the ISS railway. Jeffrey Ashby, Pamela Melroy, Sandra Magnus, David Wolf, Piers Sellers and Fyodor Yurchikhin crewed the flight.

### 9 Sep 2006
### STS-115: Atlantis

Launch was delayed by a lightning strike at the launchpad, which signalled the arrival of Tropical Storm Ernesto. Brent Jett Jr., Heidemarie Stefanyshyn-Piper, Joseph Tanner, Daniel Burbank, Christopher Ferguson and Steven MacLean crewed the Atlantis as it resumed efforts to assemble the ISS after a four year pause.

### 8 Jun 2007
### STS-117: Atlantis

Another ISS assembly flight, this time crewed by Frederick Sturckow, Lee Archambault, James Reilly II, Steven Swanson, Patrick Forrester, John Olivas and Clay Anderson.

## 7 Feb 2008
## STS-122: Atlantis

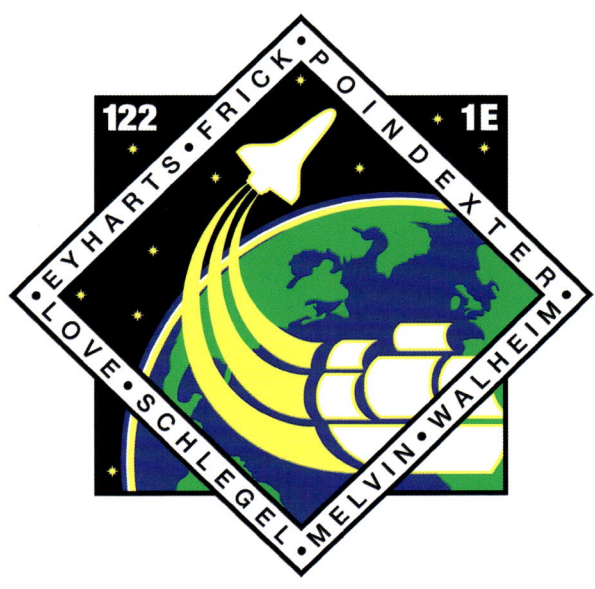

Delivery of the ESA's Columbus lab to the ISS, carried out by Steve Frick, Alan Poindexter, Rex Walheim, Stanley Love, Leland Melvin, Hans Schlegel, Léopold Eyharts and Daniel Tani.

## 11 May 2009
## STS-125: Atlantis

The final mission to repair and upgrade the Hubble Space Telescope. The crew members were Scott Altman, Gregory Johnson, Andrew Feustel, Michael Good, John Grunsfeld, Michael Massimino and Megan McArthur.

## 16 Nov 2009
## STS-129: Atlantis

This difficult supply mission to the ISS was crewed by Charles Hobaugh, Barry Wilmore, Leland Melvin, Mike Foreman, Robert Satcher Jr. and Randy Bresnik. Once the Atlantis had docked and the hatches were open, Nicole Stott left the ISS crew (Expedition 21) to join the STS-129 crew for her return to Earth, making her the last NASA astronaut to join or leave the space station via a shuttle.

## 14 May 2010
## STS-132: Atlantis

Ken Ham, Tony Antonelli, Garratt Reisman, Michael Good, Piers Sellers and Steve Bowen crewed this mission to the ISS, which was initially scheduled to be Atlantis' final flight. That honour would, however, go to STS-135.

## 8 Jul 2011
## STS-135: Atlantis

The space shuttle programme's final mission. Crewed by Christopher Ferguson, Douglas Hurley, Sandra Magnus and Rex Walheim, nearly a million people gathered along the Florida coast to witness the last ever launch of a space shuttle. Commander Ferguson said upon landing, 'Although we got to take the ride, we sure hope that everybody who has ever worked on, or touched, or looked at, or envied or admired a space shuttle was able to take just a little part of the journey with us.'

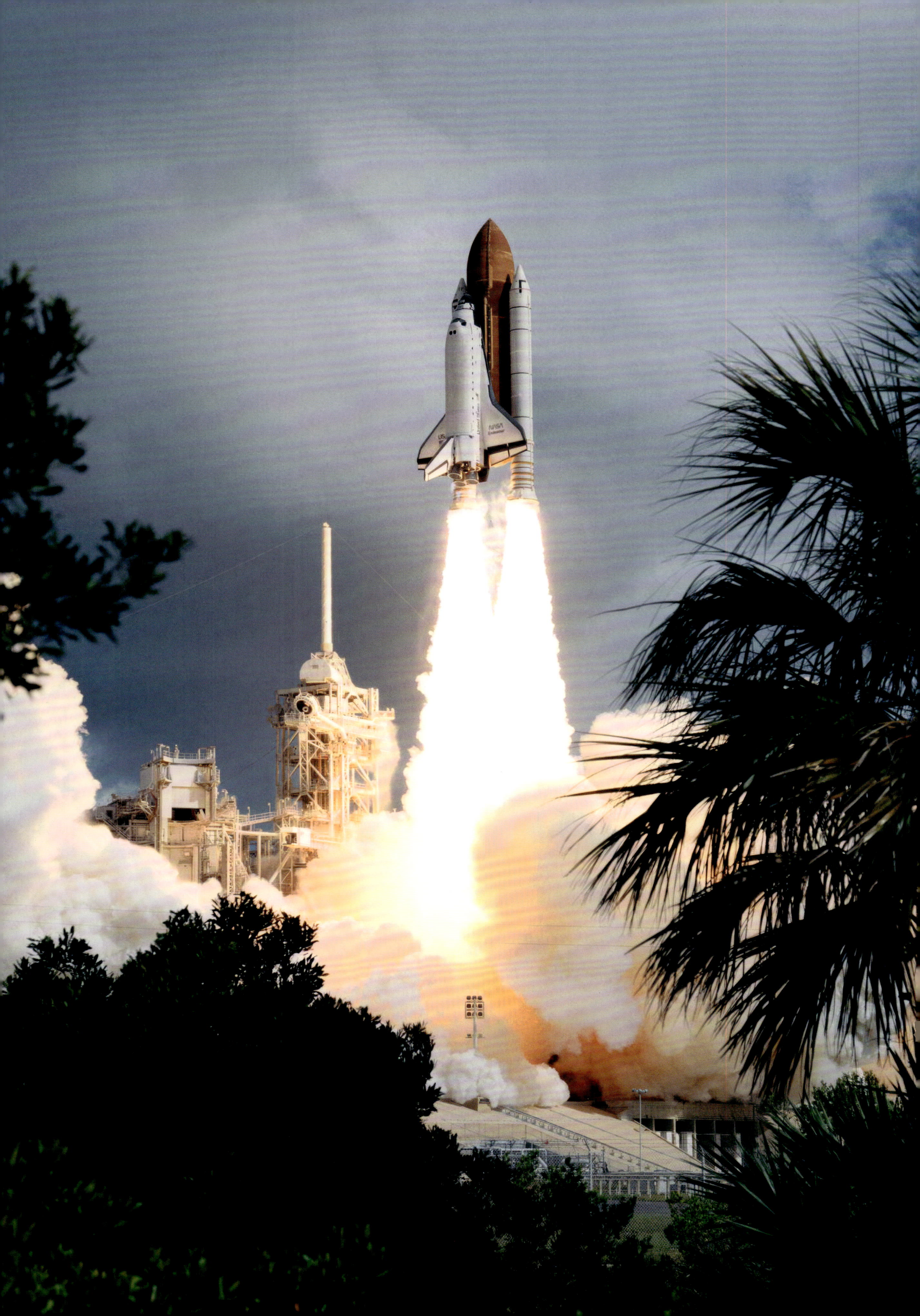

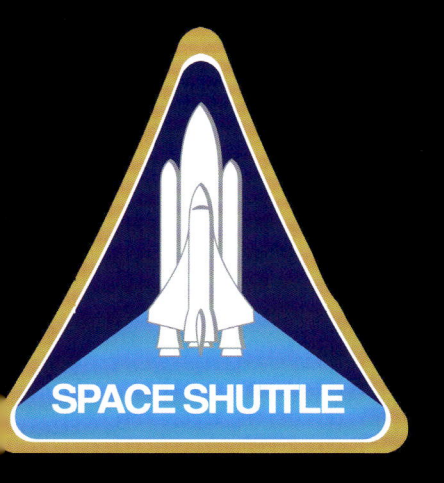

# Space Shuttle Endeavour 1992-2011

Endeavour was the fifth and final orbiter to join the space shuttle fleet. Replacing the lost Challenger, Endeavour completed a total of 25 missions, became the first shuttle to service the ISS and survived to the end of the space shuttle programme, flying for the final time in May 2011.

## 7 May 1992
### STS-49: Endeavour

The maiden flight of Endeavour. The crew of Daniel Brandenstein, Kevin Chilton, Pierre Thuot, Kathryn Thornton, Richard Hieb, Thomas Akers and Bruce Melnick recaptured the Intelsat VI, which had been stranded since its launch in 1990. They carried out repairs and relaunched it into a functional orbit.

## 12 Sep 1992
### STS-47: Endeavour

Spacelab-J was a joint NASA and NASDA (Japanese space agency) mission focused on microgravity and life sciences. The latter involved tests carried out on the crew of Robert Gibson, Curtis Brown Jr., Mark Lee, Jan Davis, Jay Apt, Mae Jemison and Mamoru Mohri, several Japanese Koi carp, chicken embryos, fruit flies, fungi seeds, frogs and frogspawn. It was the first time that a Japanese astronaut (Mohri), an African American woman (Jemison) and a married couple (Davis and Lee - the first and only) had flown on a shuttle mission.

## 13 Jan 1993
### STS-54: Endeavour

Mission to deploy the TDRS-F with crew members John Casper, Donald McMonagle, Mario Runco, Gregory Harbaugh and Susan Helms.

## 21 Jun 1993
### STS-57: Endeavour

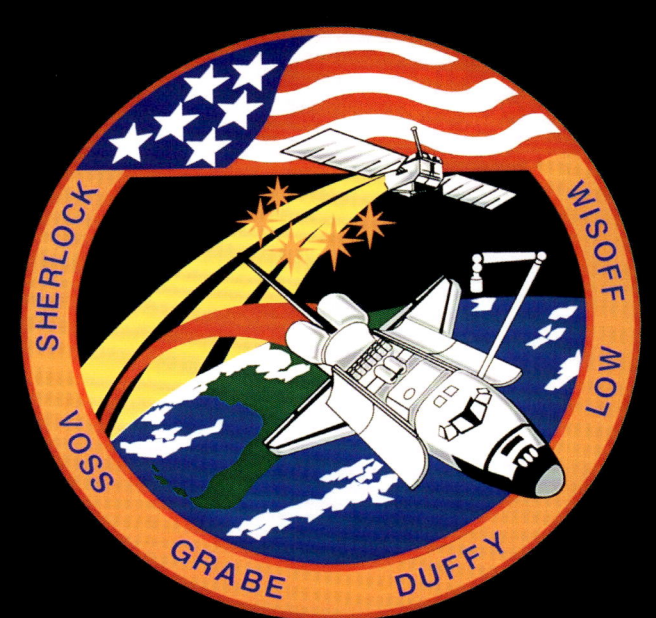

The first flight of Spacehab - a large, commercially developed, pressurised laboratory. Launch was rescheduled several times, including because a loud noise was heard after the shuttle arrived at the pad (later attributed to a ball joint inside a hydrogen tank). The crew were Ronald Grabe, Brian Duffy, David Low, Nancy Sherlock, Peter Wisoff and Janice Voss.

## 2 Dec 1993
## STS-61: Endeavour

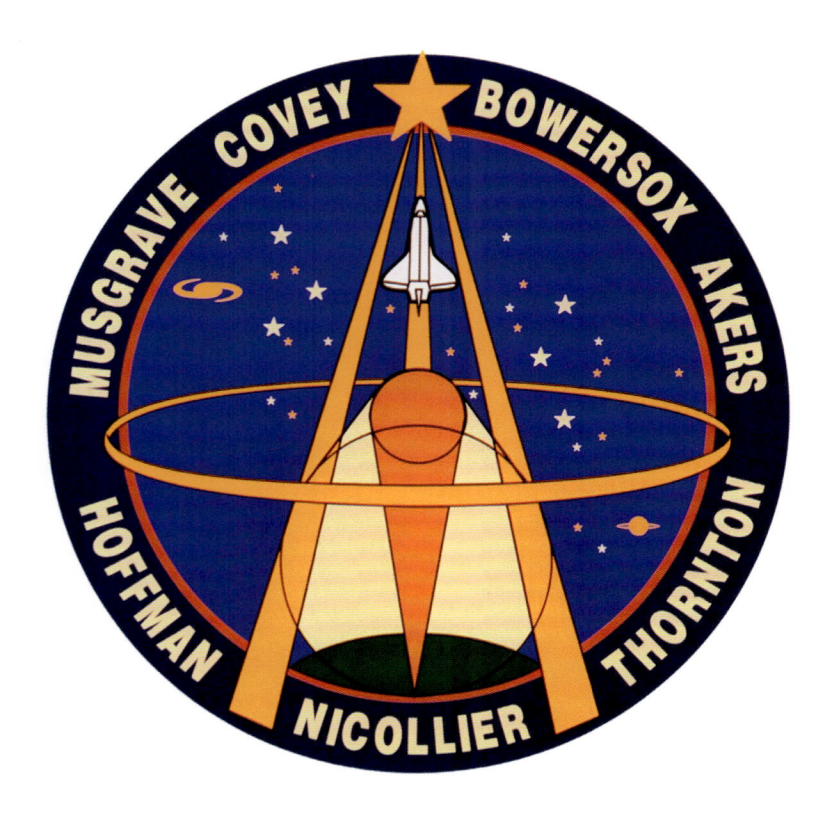

The first mission to service the Hubble Space Telescope. This complicated work was carried out by a crew of Richard Covey, Kenneth Bowersox, Story Musgrave, Kathryn Thornton, Claude Nicollier, Jeffrey Hoffman and Tom Akers. Thornton set a new US record for the longest spacewalk of 29 hours and 39 minutes.

## 9 Apr 1994
## STS-59: Endeavour

The deployment of the Space Radar Laboratory (SRL-1). Sidney Gutierrez, Kevin Chilton, Linda Godwin, Jay Apt, Michael Clifford and Thomas Jones crewed the flight.

## 30 Sep 1994
## STS-68: Endeavour

Second flight of the SLR. A handful of unusual occurrences were captured, including the eruption Klyuchevskaya Sopka, the tallest active volcano in Eurasia. The crew members were Michael Baker, Terrence Wilcutt, Thomas Jones, Steven Smith, Daniel Bursch and Peter Wisoff.

## 2 Mar 1995
## STS-67: Endeavour

Second flight of the Astro Observatory. Lasting 16 days, this was a mission to peer into the far ultraviolet depths of space. Stephen Oswald, William Gregory, Tamara Jernigan, John Grunsfeld, Wendy Lawrence, Ronald Parise and Samuel Durrance crewed the flight.

### 7 Sep 1995
### STS-69: Endeavour

This was the first mission to deploy and retrieve two separate payloads in the same flight. The Spartan 201-03 was designed to study the Sun's outer atmosphere and the solar winds that flow from it, in tandem with the passage of Ulysses spacecraft. The Wake Shield Facility (WSF-2) was the first spacecraft ever to fly away from the shuttle rather than the shuttle flying away from the craft. The crew members were David Walker, Kenneth Cockrell, James Voss, James Newman and Michael Gernhardt.

### 11 Jan 1996
### STS-72: Endeavour

Brian Duffy, Brent Jett, Leroy Chiao, Daniel Barry, Winston Scott and Koichi Wakata flew this mission to retrieve a Japanese satellite and deploy and retrieve a NASA payload.

### 19 May 1996
### STS-77: Endeavour

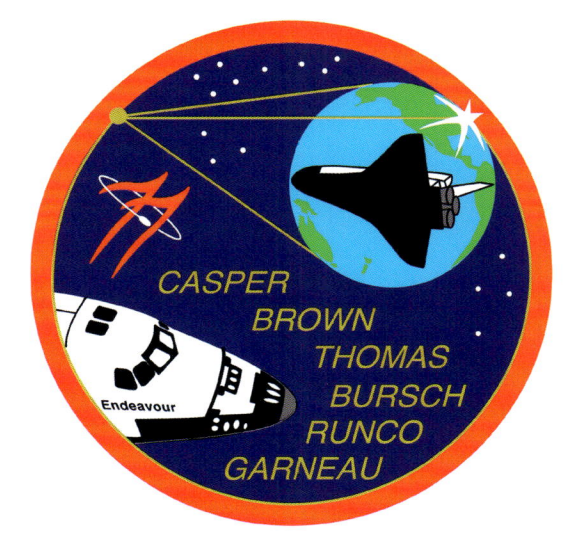

John Casper, Curtis Brown, Daniel Bursch, Mario Runco Jr., Marc Garneau and Andrew Thomas were joined by a variety of aquatic species, such as starfish, mussels and sea urchins for this 10-day research mission.

### 22 Jan 1998
### STS-89: Endeavour

The delivery of cosmonaut Salizhan Sharipov and astronaut Andy Thomas to Mir. Thomas replaced David Wolf onboard the station, but only after swapping Sokol spacesuits when Thomas found his didn't fit. Their Endeavour crewmates were Terrence Wilcutt, Joe Edwards Jr., Bonnie Dunbar, James Reilly and Michael Anderson.

### 4 Dec 1998
### STS-88: Endeavour

The first shuttle mission to the ISS. Robert Cabana, Frederick Sturckow, Jerry Ross, Nancy Currie, James Newman and Sergei Krikalev began the assembly of the station by attaching the American Unity module to the Russian Zarya control module. Jerry Ross set a new record by completing his seventh spacewalk.

### 11 Feb 2000
### STS-99: Endeavour

Kevin Kregel, Dominic Gorie, Mamoru Mohri, Gerhard Thiele, Janice Voss and Janet Kavandi crewed this successful Shuttle Radar Topography mission.

49

### 30 Nov 2000
### STS-97: Endeavour

Astronauts Brent Jett Jr., Michael Bloomfield, Marc Garneau, Carlos Noriega and Joseph Tanner delivered and connected the first US-supplied solar arrays to the ISS.

### 19 Apr 2001
### STS-100: Endeavour

A robotic arm, called Canadarm2, was delivered to the ISS. After successfully fixing itself to the station, it was directed to pass its own launch cradle back to Endeavour's robot arm, marking the first robot-to-robot transfer of equipment in space. Kent Rominger, Jeffrey Ashby, Yuri Lonchakov, Scott Parazynski, Umberto Guidoni, Chris Hadfield and John Phillips crewed the Endeavour.

### 5 Dec 2001
### STS-108: Endeavour

The Expedition 4 crew of Yury Onufrienko, Carl Walz and Daniel Bursch were transported to the ISS by Dominic Gorie, Mark Kelly, Linda Godwin and Daniel Tani aboard the Endeavour. The Expedition 3 crew of Frank Culbertson, Mikhail Tyurin and Vladimir Dezhurov were brought home.

### 5 Jun 2002
### STS-111: Endeavour

Kenneth Cockrell, Paul Lockhart, Philippe Perrin and Franklin Chang-Díaz transported the Expedition 5 crew – Valery Korzun, Peggy Whitson and Sergei Treschev – to the ISS and brought the Expedition 4 crew home.

### 23 Nov 2002
### STS-113: Endeavour

James Wetherbee, Paul Lockhart, Michael Lopez-Alegria and John Herrington took the Expedition 6 crew of Kenneth Bowersox, Nikolai Budarin and Don Pettit to the ISS and brought back the Expedition 5 crew.

### 8 Aug 2007
### STS-118: Endeavour

The 22nd shuttle flight to the ISS. The Endeavour crew included former elementary-school teacher Barbara Morgan, as well as Scott Kelly, Charlie Hobaugh, Dave Williams, Rick Mastracchio, Tracy Caldwell-Dyson and Alvin Drew.

### 11 Mar 2008
### STS-123: Endeavour

Dominic Gorie, Gregory Johnson, Richard Linnehan, Robert Behnken, Michael Foreman and Takao Doi delivered a Japanese logistics module and a Canadian robotics system to the ISS, along with Garrett Reisman, who joined the ISS crew upon arrival, and Leopold Eyharts, who joined the Endeavour crew for his return to Earth.

### 14 Nov 2008
### STS-126: Endeavour

Equipment delivery to the ISS, carried out by Chris Ferguson, Eric Boe, Steven Bowen, Shane Kimbrough, Heidemarie Stefanyshyn-Piper, Donald Pettit, Sandra Magnus and Gregory Chamitoff.

### 15 Jul 2009
### STS-127: Endeavour

Mark Polansky, Doug Hurley, Dave Wolf, Christopher Cassidy, Julie Payette, Tom Marshburn and Tim Kopra delivered a Japanese laboratory to the ISS.

### 8 Feb 2010
### STS-130: Endeavour

Delivery of the Tranquility module to the ISS, carried out by George Zamka, Terry Virts, Nicholas Patrick, Robert Behnken, Kathryn Hire and Stephen Robinson.

### 16 May 2011
### STS-134: Endeavour

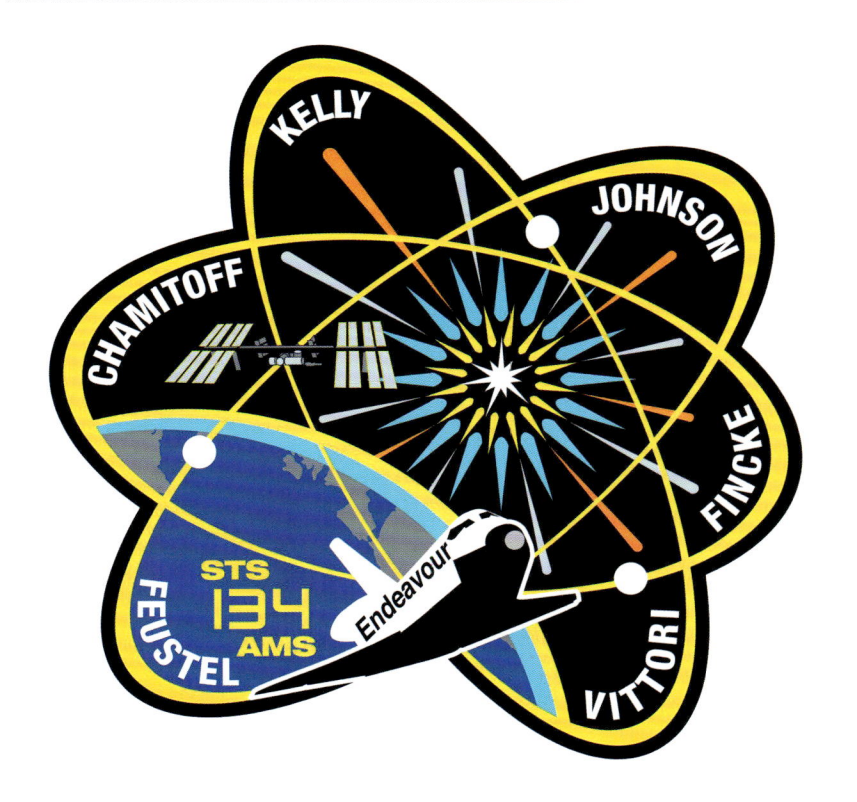

This was the final flight of NASA's youngest shuttle. For its last ever journey, Endeavour delivered a number of parts and supplies to the ISS and was crewed by Mark Kelly, Gregory Johnson, Michael Fincke, Greg Chamitoff, Andrew Feustel and Roberto Vittori.

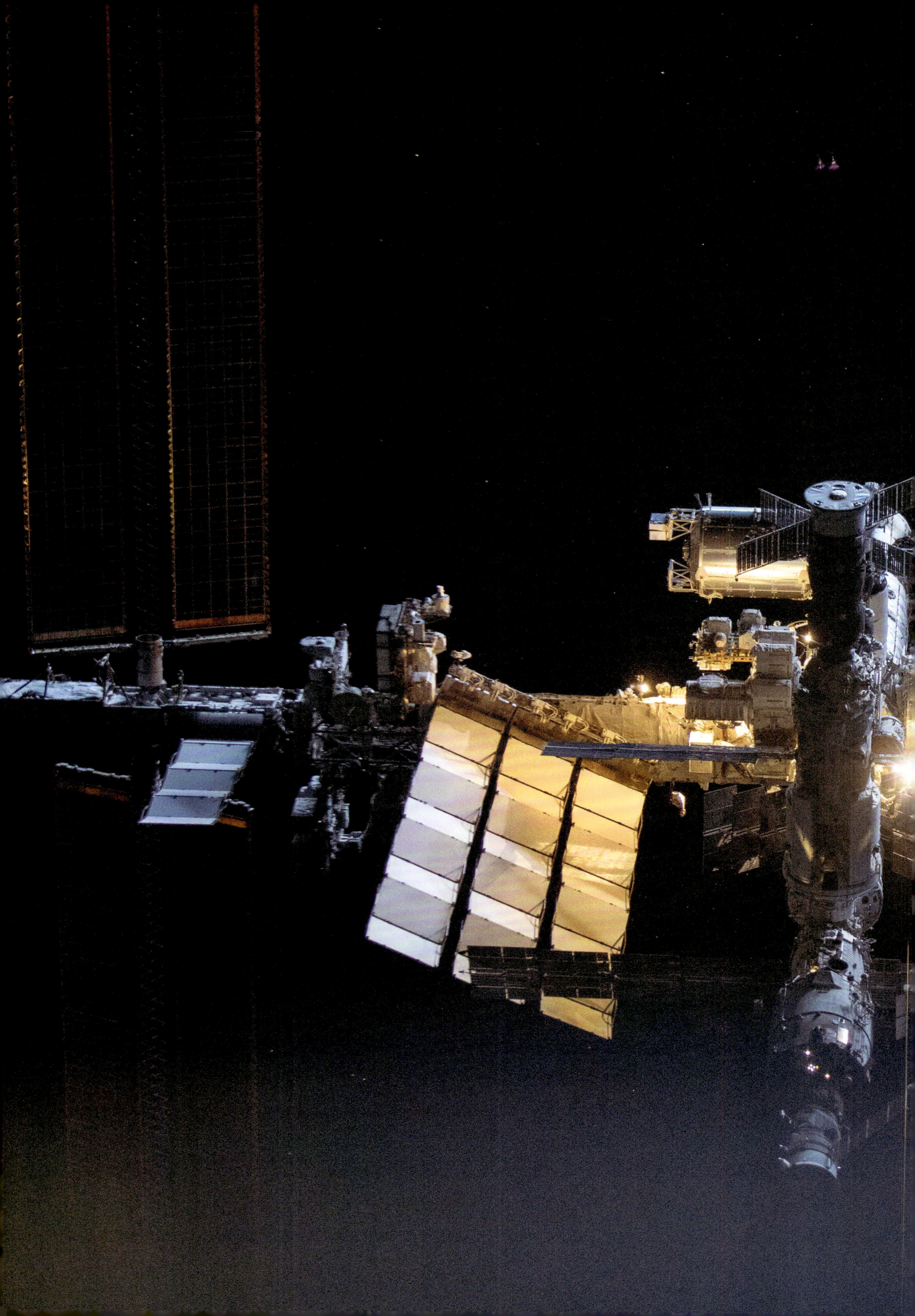

# Chapter 3: International Space Station

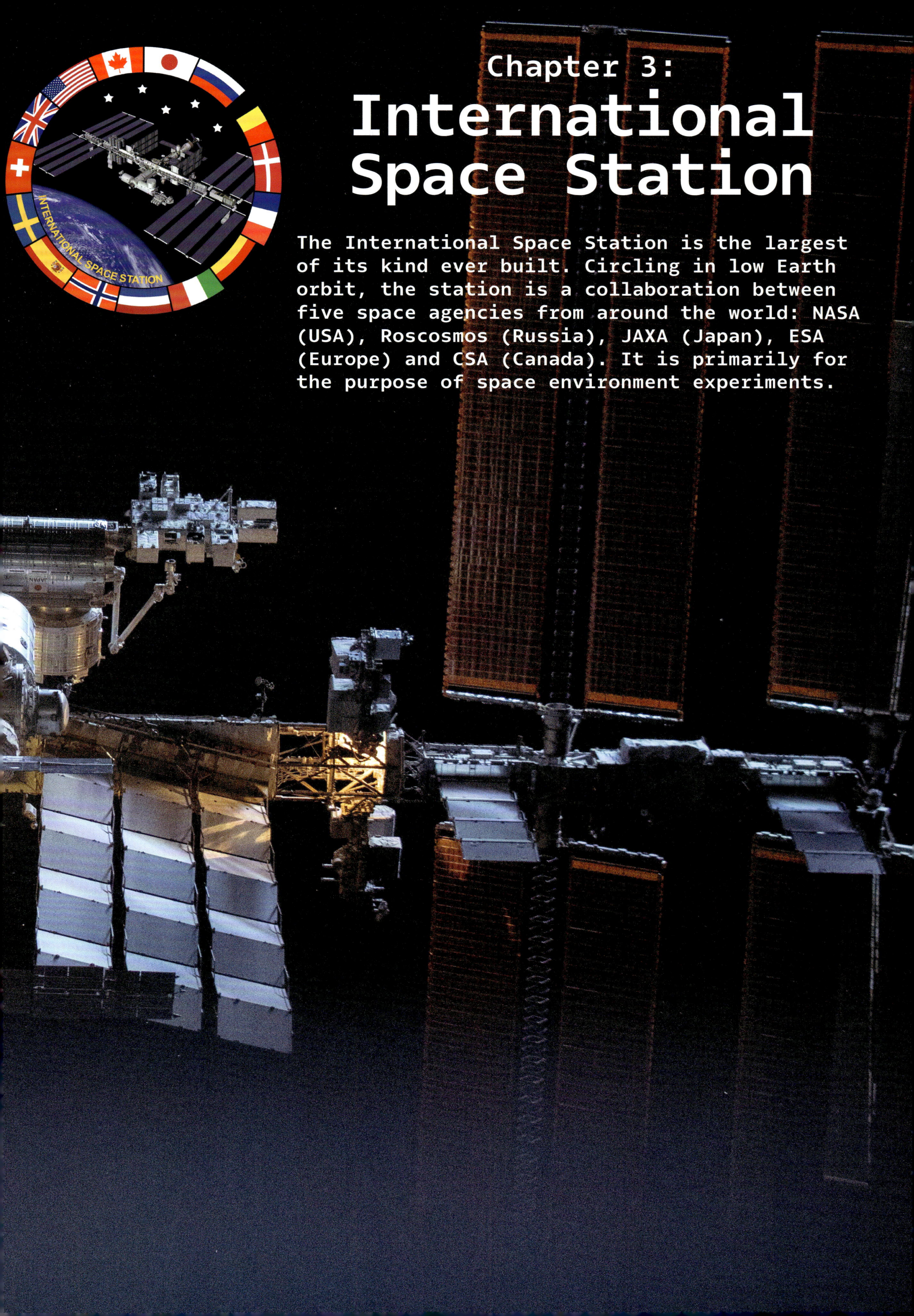

The International Space Station is the largest of its kind ever built. Circling in low Earth orbit, the station is a collaboration between five space agencies from around the world: NASA (USA), Roscosmos (Russia), JAXA (Japan), ESA (Europe) and CSA (Canada). It is primarily for the purpose of space environment experiments.

| ISS Expedition 1<br>2 Nov 2000 – 21 Mar 2001 | ISS Expedition 2<br>8 Mar – 22 Aug 2001 | ISS Expedition 3<br>10 Aug – 17 Dec 2001 |
|---|---|---|
| ISS Expedition 4<br>5 Dec 2001 – 19 Jun 2002 | ISS Expedition 5<br>5 Jun – 7 Dec 2002 |  ISS Expedition 6<br>24 Nov 2002 – 4 May 2003 |
| ISS Expedition 7<br>26 Apr – 28 Oct 2003 | ISS Expedition 8<br>18 Oct 2003 – 30 Apr 2004 | ISS Expedition 9<br>19 Apr – 24 Oct 2004 |
| ISS Expedition 10<br>16 Oct 2004 – 24 Apr 2005 | ISS Expedition 11<br>15 Apr – 11 Oct 2005 | ISS Expedition 12<br>1 Oct 2005 – 8 Apr 2006 |

| ISS Expedition 13<br>30 Mar – 28 Sep 2006 | ISS Expedition 14<br>18 Sep 2006 – 21 Apr 2007 |  ISS Expedition 15<br>7 Apr – 21 Oct 2007 |
|---|---|---|
| ISS Expedition 16<br>10 Oct 2007 – 19 Apr 2008 | ISS Expedition 17<br>8 Apr – 24 Oct 2008 | ISS Expedition 18<br>12 Oct 2008 – 8 Apr 2009 |
| ISS Expedition 19<br>26 Mar – 29 May 2009 | ISS Expedition 20<br>27 May – 11 Oct 2009 | ISS Expedition 21<br>30 Sep – 1 Dec 2009 |
|  ISS Expedition 22<br>20 Dec 2009 – 18 Mar 2010 | ISS Expedition 23<br>18 Mar – 2 Jun 2010 | ISS Expedition 24<br>2 Jun – 27 Sep 2010 |

| | | |
|---|---|---|
| **ISS Expedition 25**<br>25 Sep – 26 Nov 2010 | **ISS Expedition 26**<br>26 Nov 2010 – 16 Mar 2011 | **ISS Expedition 27**<br>16 Mar – 23 May 2011 |
| **ISS Expedition 28**<br>23 May – 16 Sep 2011 | **ISS Expedition 29**<br>16 Sep – 21 Nov 2011 | **ISS Expedition 30**<br>21 Nov 2011 – 27 Apr 2012 |
| **ISS Expedition 31**<br>27 Apr – 1 Jul 2012 |  **ISS Expedition 32**<br>1 Jul – 16 Sep 2012 | **ISS Expedition 33**<br>16 Sep – 18 Nov 2012 |
| **ISS Expedition 34**<br>18 Nov 2012 – 15 Mar 2013 | **ISS Expedition 35**<br>15 Mar – 13 May 2013 | **ISS Expedition 36**<br>13 May – 10 Sep 2013 |

| ISS Expedition 37<br>10 Sep 2013 - 10 Nov 2013 | ISS Expedition 38<br>10 Nov 2013 - 11 Mar 2014 | ISS Expedition 39<br>11 Mar - 13 May 2014 |
|---|---|---|
| ISS Expedition 40<br>13 May - 10 Sep 2014 | ISS Expedition 41<br>10 Sep - 10 Nov 2014 | ISS Expedition 42<br>10 Nov 2014 - 11 Mar 2015 |
| ISS Expedition 43<br>11 Mar - 11 Jun 2015 | ISS Expedition 44<br>11 Jun - 11 Sep 2015 | ISS Expedition 45<br>11 Sep - 11 Dec 2014 |
| ISS Expedition 46<br>11 Dec 2015 - 2 Mar 2016 | ISS Expedition 47<br>2 Mar - 18 Jun 2016 | ISS Expedition 48<br>18 Jun - 6 Sep 2016 |

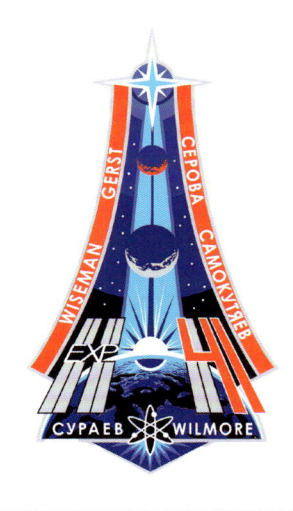

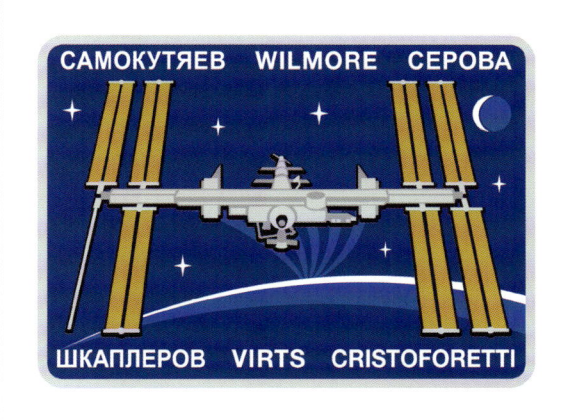

| ISS Expedition 49 | ISS Expedition 50 | ISS Expedition 51 |
| --- | --- | --- |
| 6 Sep - 30 Oct 2016 | 30 Oct 2016 - 10 Apr 2017 | 10 Apr - 2 Jun 2017 |
| ISS Expedition 52 | ISS Expedition 53 | ISS Expedition 54 |
| 2 Jun - 2 Sep 2017 | 2 Sep - 14 Dec 2017 | 14 Dec 2017 - 27 Feb 2018 |
| ISS Expedition 55 | ISS Expedition 56 | ISS Expedition 57 |
| 27 Feb - 3 Jun 2018 | 3 Jun - 4 Oct 2018 | 4 Oct - 20 Dec 2018 |
| ISS Expedition 58 | ISS Expedition 59 | ISS Expedition 60 |
| 20 Dec 2018 - 15 Mar 2019 | 15 Mar - 24 Jun 2019 | 4 Jun - 3 Oct 2019 |

## ISS Expedition 61
3 Oct 2019 – 6 Feb 2020

## ISS Expedition 62
6 Feb – 17 Apr 2020

## ISS Expedition 63
17 Apr – 21 Oct 2020

## ISS Expedition 64
21 Oct 2020 – 17 Apr 2021

## ISS Expedition 65
17 Apr – 17 Oct 2021

## ISS Expedition 66
17 Oct 2021 – 30 Mar 2022

## ISS Expedition 67
30 Mar – 29 Sep 2022

## ISS Expedition 68
29 Sep 2022 – 28 Mar 2023

## ISS Expedition 69
28 Mar – 27 Sep 2023

## ISS Expedition 70
27 Sep 2023 – Apr 5 2024

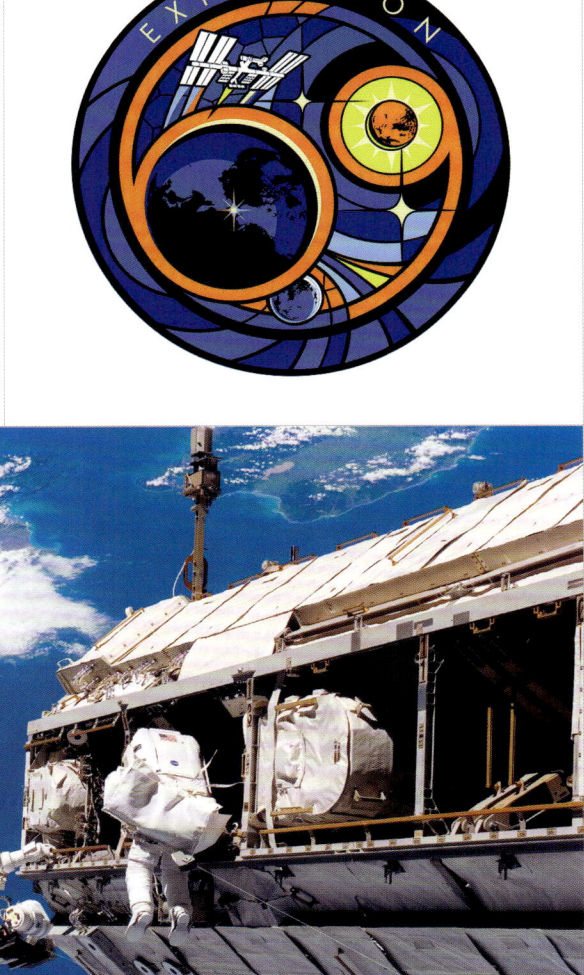

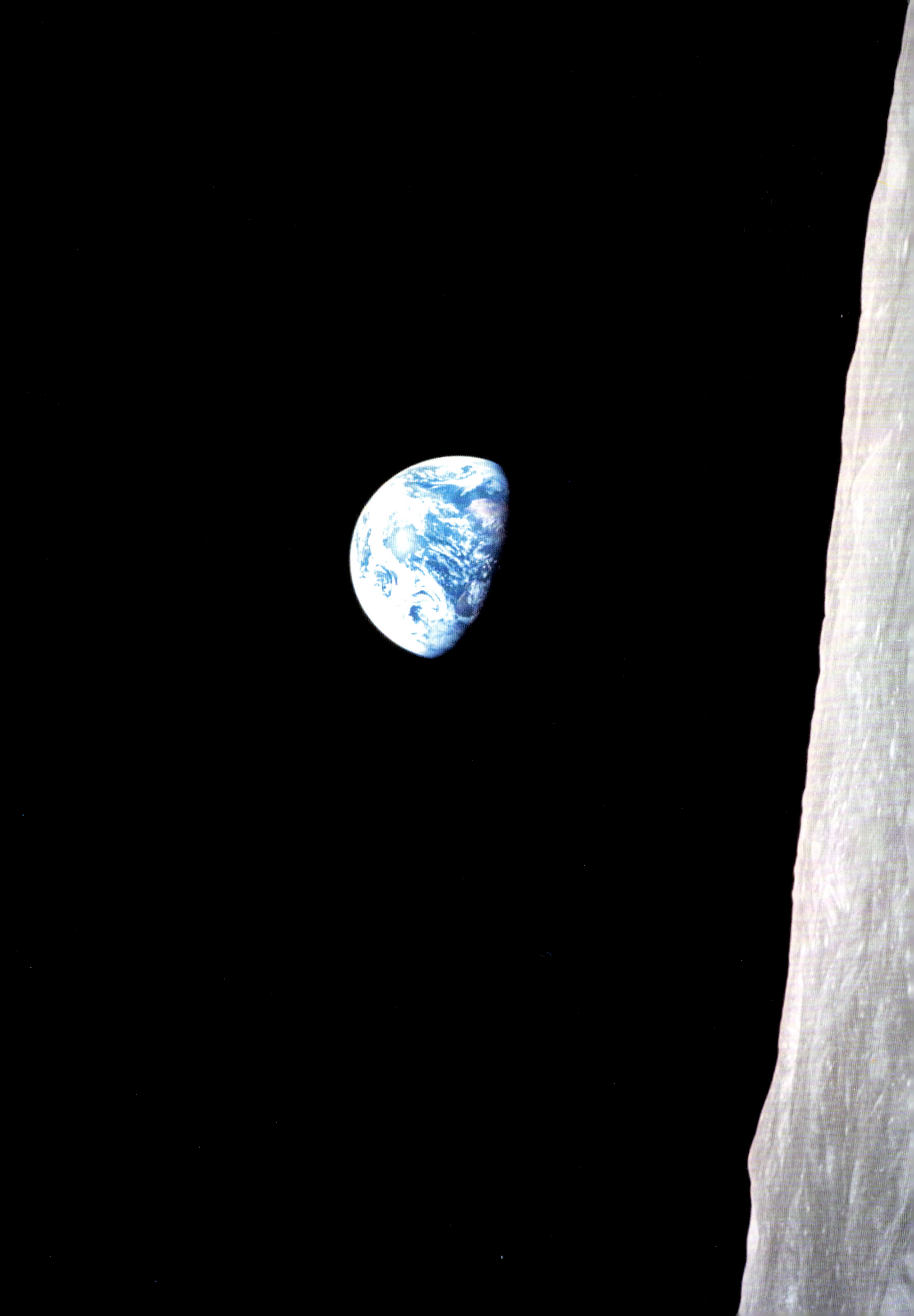

# Chapter 4:
# The Future of Spaceflight

*SpaceX, Artemis*

| | |
|---|---|
| **7 October 2012** | The launch of the first commercial resupply mission to the ISS, which was undertaken by the SpaceX Dragon, nicknamed the Endeavour. |
| **11 December 2017** | The new Artemis programme was announced. |
| **25 December 2021** | NASA launched the powerful James Webb Space Telescope, allowing humanity to see further into the cosmos than ever before. |
| **16 November 2022** | Artemis I. The first flight of the Artemis programme saw an uncrewed Orion spacecraft complete a lunar orbit before splashing down safely off the coast of Baja California, Mexico. |

### 2020 SpaceX Demo-2

Robert Behnken and Doug Hurley flew this mission to test and validate the SpaceX human transport systems for use in future NASA programmes. They named this Dragon vehicle 'Endeavour' after the shuttle that first flew them both into space.

### 2022 Artemis 1

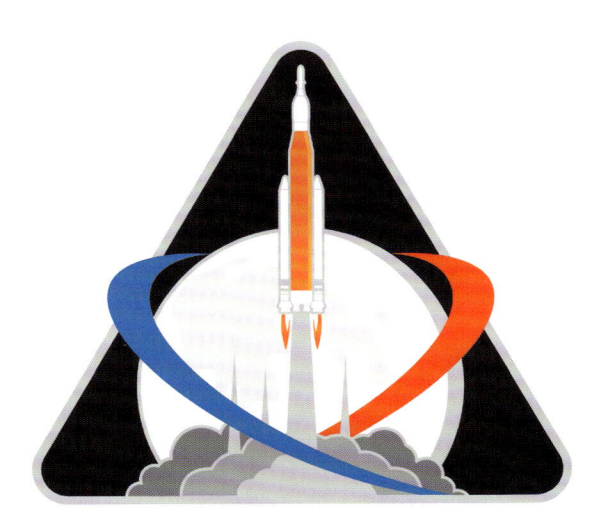

The first test flight of the Orion spacecraft and SLS rocket. Uncrewed, Orion flew far beyond the Moon before circling back to Earth.

# SpaceX
## 2020-ongoing

Founded in 2002 by Elon Musk, SpaceX has been credited with launching an era of commercial spaceflights. It was the first private company to successfully dock a crewed spacecraft with the ISS. The company's aims include establishing a colony on Mars and carrying humans to other destinations in the solar system.

### 2020 SpaceX Crew 1

The first long-duration mission of the Commercial Crew programme saw NASA's Mike Hopkins, Shannon Walker, Victor Glover and JAXA's Soichi Noguchi fly SpaceX's Dragon spacecraft to the ISS.

### 2020 SpaceX Crew 2

The second ISS crew flight using the Dragon. Astronauts Shane Kimbrough, Megan McArthur, Thomas Pesquet and Akihiko Hoshide conducted various studies once aboard the ISS, including a test of cotton growth in microgravity in an attempt to identify varieties that may require less water and fewer pesticides to thrive.

### 2021 SpaceX Crew 3

The Commercial Crew programme established a tradition of the crew naming their spacecraft and, for this third mission (and fifth flight overall), Raja Chari, Thomas Marshburn, Matthias Maurer and Kayla Barron named their vehicle 'Endurance' in honour of the NASA and SpaceX teams who had endured throughout the COVID-19 pandemic to complete the manufacture of the craft and train the astronauts who flew it to the ISS.

### 2022 SpaceX Crew 4

Another new Dragon craft, this time named 'Freedom', was flown to the ISS, carrying Kjell Lindgren, Bob Hines, Jessica Watkins and Samantha Cristoforetti.

### 2022 SpaceX Crew 5

The second mission for the SpaceX Dragon Endurance, ferrying Nicole Anapu Mann, Josh Cassada, Koichi Wakata and Anna Kikina to the ISS.

### 2023 SpaceX Crew 6

A United Arab Emirates astronaut, Sultan Al Neyadi, joined Stephen Bowen and Warren Hoburg from NASA and Andrey Fedyaev from Russia aboard this flight to the ISS.

### 2023 SpaceX Crew 7

A fully multinational mission to the ISS, the Dragon Endurance successfully delivered NASA's Jasmin Moghbeli, ESA's Andreas Mogensen, JAXA's Satoshi Furukawa and Roscosmos's Konstantin Borisov to the station.

### 2024 SpaceX Crew 8

A fourth flight for the SpaceX Dragon Endeavour, transporting Matthew Dominick, Michael Barratt, Jeanette Epps and Alexander Grebenkin to the ISS.

# Artemis Programme
## 2017-ongoing

NASA's Artemis programme has the dual purpose of sending humanity back to the Moon to establish our first long-term presence on the lunar surface, while simultaneously advancing plans for the eagerly anticipated missions to Mars.

ARTEMIS II
FOR ALL HUMANITY

**Picture Credits:**
Page 4: Dennis Hallinan/Alamy Stock Photo
Page 6: Camerique/Alamy Stock Photo
Page 8: Jim McDivitt/NASA/Roger Ressmeyer/Corbis/VCG via Getty Images
Page 12: Central Press/Hulton Archive/Getty Images
Page 17: NASA Image Collection/Alamy Stock Photo
Page 29 (bottom-right): NASA Photo/Alamy Stock Photo
Page 30: Space Frontiers/Archive Photos/Hulton Archive/Getty Images
Page 44: Science History Images/Alamy Stock Photo
Page 59: LWM/NASA/LANDSAT/Alamy Stock Photo
Page 60: American Photo Archive/ Alamy Stock Photo
All other images courtesy of NASA

Back cover image of astronaut Sally Ride's flight jacket: Heritage Art/Heritage Images via Getty Images

© 2025 Bill Schwartz
World copyright reserved

ISBN: 978 1 78884 289 1

The right of Bill Schwartz to be identified as author of this work has been asserted by him in accordance with the Copyright, Designs and Patents Act 1988

All rights reserved. No part of this publication may be reproduced, stored in a retrieval system, or transmitted in any form or by any means electronic, mechanical, photocopying, recording or otherwise, without the prior permission of the publisher

A CIP record for this book is available from the British Library

The author and publisher gratefully acknowledge the permission granted to reproduce the copyright material in this book. Every effort has been made to trace copyright holders and to obtain their permission for the use of copyright material. The publisher apologises for any errors or omissions in the text and would be grateful if notified of any corrections that should be incorporated in future reprints or editions of this book.

Editors: Alice Bowden and Stewart Norvill
Designer: Stephen Farrow
Reprographics: Corban Wilkin

Printed in China
for ACC Art Books Ltd., Woodbridge, Suffolk, UK

www.accartbooks.com

**ACC ART BOOKS**